I0468033

Hydrogeologic Framework and Groundwater/ Surface-Water Interactions of the Chehalis River Basin, Southwestern Washington

By Andrew S. Gendaszek

Prepared in cooperation with the U.S. Army Corps of Engineers, Washington State Department of Ecology, and the Chehalis Basin Partnership

Scientific Investigations Report 2011–5160

U.S. Department of the Interior
U.S. Geological Survey

U.S. Department of the Interior
KEN SALAZAR, Secretary

U.S. Geological Survey
Marcia K. McNutt, Director

U.S. Geological Survey, Reston, Virginia: 2011

For more information on the USGS—the Federal source for science about the Earth, its natural and living resources, natural hazards, and the environment, visit http://www.usgs.gov or call 1–888–ASK–USGS.

For an overview of USGS information products, including maps, imagery, and publications, visit http://www.usgs.gov/pubprod

To order this and other USGS information products, visit http://store.usgs.gov

Suggested citation:
Gendaszek, A.S., 2011, Hydrogeologic framework and groundwater/surface-water interactions of the Chehalis River basin, Washington: U.S. Geological Survey Scientific Investigations Report 2011-5160, 42 p.

Contents

Plate

Plate 1. Map and hydrogeologic sections showing location of inventoried wells, surficial geology, and hydrogeologic units in the Chehalis River basin, Washington.

Figures

Tables

Conversion Factors and Datums

Inch/Pound to SI

Multiply	By	To obtain
Length		
foot (ft)	0.3048	meter (m)
mile (mi)	1.609	kilometer (km)
Area		
square mile (mi2)	2.590	square kilometer (km^2)
Flow Rate		
cubic foot per day (ft^3/d)	0.02832	cubic meter per day (m^3/d)
cubic foot per second (ft^3/s)	0.02832	cubic meter per second (m^3/s)
cubic foot per second per square mile [(ft^3/s)/mi^2]	0.01093	cubic meter per second per square kilometer [(m^3/s)/km^2]
cubic foot per mile (ft^3/mi)	0.01760	cubic meters per kilometer (m^3/km)
inches per year (in/yr)	2.54	centimeters per year (cm/yr)
foot per day (ft/d)	0.3048	meter per day (m/d)
foot per mile (ft/mi)	0.1894	meter per kilometer (m/km)

Temperature in degrees Celsius (°C) may be converted to degrees Fahrenheit (°F) as follows:

$$°F=(1.8×°C)+32.$$

Temperature in degrees Fahrenheit (°F) may be converted to degrees Celsius (°C) as follows:

$$°C=(°F-32)/1.8.$$

Datums

Vertical coordinate information is referenced to the insert datum name (and abbreviation) here for instance, "North American Vertical Datum of 1988 (NAVD 88)."

Horizontal coordinate information is referenced to the insert datum name (and abbreviation) here for instance, "North American Datum of 1983 (NAD 83)."

Altitude, as used in this report, refers to distance above the vertical datum.

*Transmissivity: The standard unit for transmissivity is cubic foot per day per square foot times foot of aquifer thickness [(ft^3/d)/ft^2]ft. In this report, the mathematically reduced form, foot squared per day (ft^2/d), is used for convenience.

Abbreviations and Acronyms

ADCP	Acoustic Doppler Current Profiler
BDRK	Bedrock Low Permeability Unit
CBP	Chehalis Basin Partnership
Kya	Thousand Years Ago
NWIS	National Water Information System
USGS	U.S. Geological Survey
WADOE	Washington State Department of Ecology

Well-Numbering System

In Washington, wells are assigned numbers that identify their location within a township, range, section, and 40-acre tract. Number 15N/02W-12A01 indicates that the well is in Township 15 North and Range 02 West, north and west of the Willamette Base Line and Meridian, respectively. The numbers immediately following the hyphen indicate the section (12) within the township; the letter following the section gives the 40-acre tract of the section, as shown on figure 4. The two-digit sequence number (01) following the letter indicates that the well was the first one inventoried by project personnel in that 40-acre tract. A "D" following the sequence number indicates that the well has been deepened. In the illustrations of this report, wells are identified individually by only the section and 40-acre tract, such as 12A01; township and range are shown on the map borders.

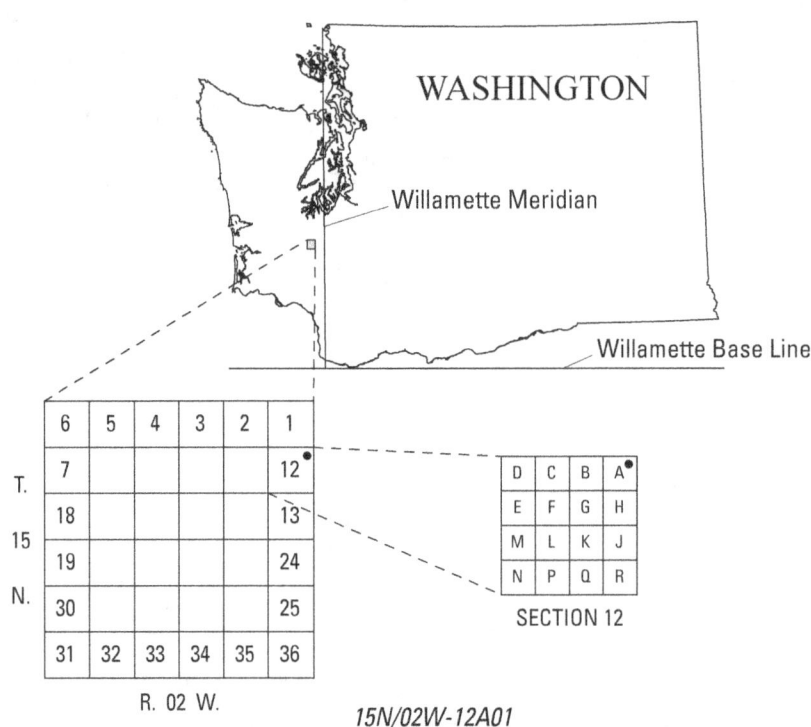

Hydrogeologic Framework and Groundwater/ Surface-Water Interactions of the Chehalis River Basin, Southwestern Washington

By Andrew S. Gendaszek

Abstract

The Chehalis River has the largest drainage basin of any river entirely contained within the State of Washington with a watershed of approximately 2,700 mi^2 and has correspondingly diverse geology and land use. Demands for water resources have prompted the local citizens and governments of the Chehalis River basin to coordinate with Federal, State and Tribal agencies through the Chehalis Basin Partnership to develop a long-term watershed management plan. The recognition of the interdependence of groundwater and surface-water resources of the Chehalis River basin became the impetus for this study, the purpose of which is to describe the hydrogeologic framework and groundwater/ surface-water interactions of the Chehalis River basin.

Surficial geologic maps and 372 drillers' lithostratigraphic logs were used to generalize the basin-wide hydrogeologic framework. Five hydrogeologic units that include aquifers within unconsolidated glacial and alluvial sediments separated by discontinuous confining units were identified. These five units are bounded by a low permeability unit comprised of Tertiary bedrock.

A water table map, and generalized groundwater-flow directions in the surficial aquifers, were delineated from water levels measured in wells between July and September 2009. Groundwater generally follows landsurface-topography from the uplands to the alluvial valley of the Chehalis River. Groundwater gradients are highest in tributary valleys such as the Newaukum River valley (approximately 23 cubic feet per mile), relatively flat in the central Chehalis River valley (approximately 6 cubic feet per mile), and become tidally influenced near the outlet of the Chehalis River to Grays Harbor.

The dynamic interaction between groundwater and surface-water was observed through the synoptic streamflow measurements, termed a seepage run, made during August 2010, and monitoring of water levels in wells during the 2010 Water Year. The seepage run revealed an overall gain of 56.8 ± 23.7 cubic feet per second over 32.8 river miles (1.7 cubic feet per second per mile), and alternating gains and losses of streamflow ranging from -48.3 to 30.9 cubic feet per second per mile, which became more pronounced on the Chehalis River downstream of Grand Mound. However, most gains and losses were within measurement error. Groundwater levels measured in wells in unconsolidated sediments fluctuated with changes in stream stage, often within several hours. These fluctuations were set by precipitation events in the upper Chehalis River basin and tides of the Pacific Ocean in the lower Chehalis River basin.

Introduction

The Chehalis River flows about 125 mi from its headwaters in the Willipa Hills through forestland, agricultural lands, the cities of Chehalis and Centralia, and the Confederated Tribes of the Chehalis Reservation to its outlet at Grays Harbor (fig. 1). Most development and water use within the Chehalis River basin is within the valleys of the Chehalis River and its major tributaries. Wells located within the major river valleys generally are completed within aquifers in the unconsolidated glacial and alluvial sediments and do not penetrate bedrock comprised of basalt and sedimentary rock.

The groundwater and surface-water systems of the Chehalis River basin supply residential, agricultural, and industrial users with water while sustaining instream and riparian ecosystems. These complex systems are connected in the Chehalis River basin (Sinclair and Hirschey, 1992; Pitz and others, 2005; Ely and others, 2008) and may not be fully understood independently of each other. Low summer streamflows and coincident high water temperatures impact the migration of anadromous fish on the Chehalis River and its tributaries, while seasonally low water levels in rivers and wells impact water availability for human use. Future water availability, particularly during the summer low-flow period, has prompted concern among the citizens and governments of the Chehalis River basin, as population growth increases the demand for water resources. Characterization of the shallow groundwater system and its interaction with the surface-water system is therefore necessary to plan for current and future water needs in the Chehalis River basin.

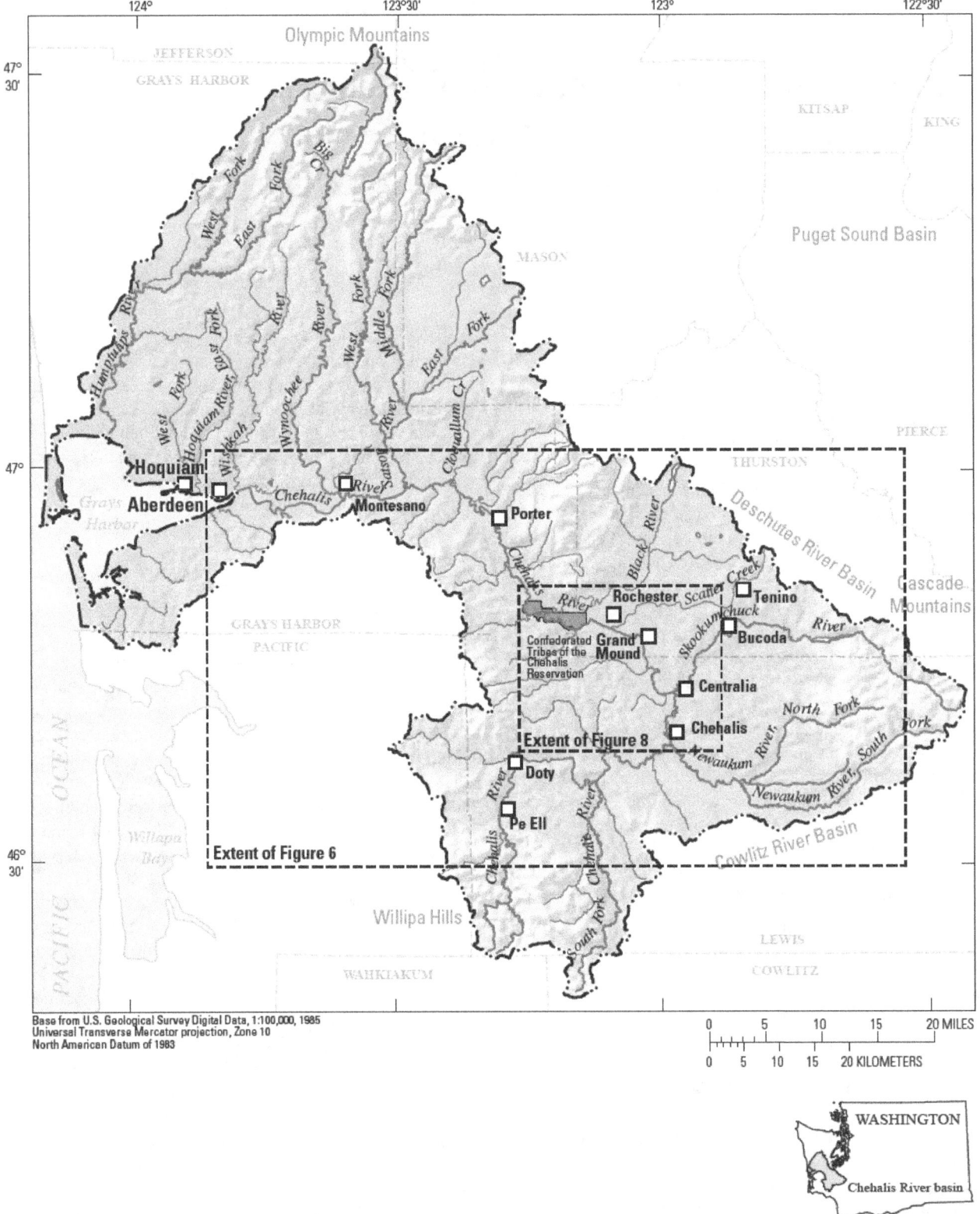

Figure 1. Location of the Chehalis River basin, southwestern Washington.

The Chehalis Basin Partnership (CBP), created by local citizens and governments, together with State and Tribal agencies acting under the guidelines of the Watershed Management Act of 1998 (Washington State Engrossed Substitute House Bill 2514), is working to implement a sustainable watershed plan that considers the groundwater and surface-water systems together, to address concerns about water quantity and quality, in-stream flows, and fish habitat.

Parts of the groundwater-flow system of the Chehalis River basin have been previously described (for example, Sinclair and Hirschey, 1992; Drost and others, 1998; Pitz and others, 2005), but a comprehensive, basin-wide assessment of the groundwater resources in the Chehalis River basin had not been done prior to this report. The basinwide hydrogeologic framework presented in this report was described in the context of generalized (Ely and others, 2008) and spatially refined information about groundwater/surface-water interactions presented in this report. This information will allow the CBP to more effectively plan and manage the water resources of the Chehalis River basin for use by humans, fish, and wildlife as population and development increases.

Purpose and Scope

The purpose of this report is to describe the generalized hydrogeologic framework of the Chehalis River basin, and to characterize the interactions between the groundwater-flow system and the Chehalis River and its major tributaries. The hydrogeologic framework presented in this report builds upon previously described hydrogeologic frameworks of parts of the Chehalis River basin (Sinclair and Hirschey, 1992; Drost and others, 1998; Pitz and others, 2005). This report includes an overview of the geologic history of the Chehalis River basin, a characterization of significant hydrogeologic units, groundwater levels, and generalized flow directions within significant aquifers. Streamflow gains and losses at selected intervals along the Chehalis River and its selected tributaries also are presented to describe groundwater/surface-water interactions in the Chehalis River basin.

Description of Study Area

The Chehalis River basin encompasses approximately 2,700 mi^2 in southwestern Washington. The basin is bounded to the north by the Olympic Mountains and a low divide with the Puget Sound basin, to the east by the Cascade Mountains,

to the south by the Willipa Hills, and to the west by the Pacific Ocean (fig. 1). Altitudes of the Chehalis River basin range from approximately 5,000 ft at Capitol Peak in the Olympic Mountains to sea level at the outlet of the Chehalis River at Grays Harbor. The Chehalis River basin is predominantly forested (74 percent) with some developed land (7 percent, primarily in the Chehalis-Centralia and Aberdeen-Hoquiam areas), wetlands (6 percent), agricultural lands (5 percent; Homer and others, 2004) in the main river valleys, and other land types (8 percent). The Chehalis River basin includes parts of Lewis, Pacific, Thurston, Mason, Grays Harbor, Cowlitz, and Jefferson Counties, and the entire Confederated Tribes of the Chehalis Reservation.

The temperate marine climate of the Chehalis River basin is characterized by cool, wet winters and warm, dry summers. Mean annual (1971–2000) precipitation ranges from more than 250 in. in the headwaters of the Wynoochee and Humptulips Rivers to 43 in. near the cities of Chehalis and Centralia (PRISM Group, 2011) (fig. 2). About 75 percent of the precipitation measured between 1971 and 2000 at the Centralia weather station, altitude 185 ft, fell between October and March (National Oceanic and Atmospheric Administration, 2010). Mean annual discharge at the downstream-most gaging station upstream of the tidal influence of Grays Harbor (Chehalis River at Porter, Washington: 12031000) for 1953–2009 was 4,049 ft^3/s. At this station for 1953–2009, mean monthly discharge was lowest in August (414 ft^3/s) and highest in January (9,520 ft^3/s) (fig. 3).

Three main types of rocks and sediments are exposed within the Chehalis River basin, including Tertiary sedimentary and volcanic rocks, Pleistocene glacial drift, andrecent alluvium (Snavely and others, 1958). Uplifted Tertiary marine and non-marine sedimentary rocks deposited on the western margin of North America comprise the bedrock foundation of the Chehalis River basin. Glacial drift from both alpine glaciers originating from the Cascade and Olympic Mountains and the Puget Lobe of the Cordilleran Ice Sheet were deposited in the Chehalis River basin at least twice during the Pleistocene. Recent alluvium overlies much of the Pleistocene glacial deposits within the valleys of the Chehalis River and its major tributaries. Significant groundwater resources within the Chehalis River basin are contained within the well-sorted sands and gravels of the unconsolidated deposits that were deposited by outwash streams and modern rivers (Wallace and Molenaar, 1961; Weigle and Foxworthy, 1962; Eddy, 1966; Noble and Wallace, 1966).

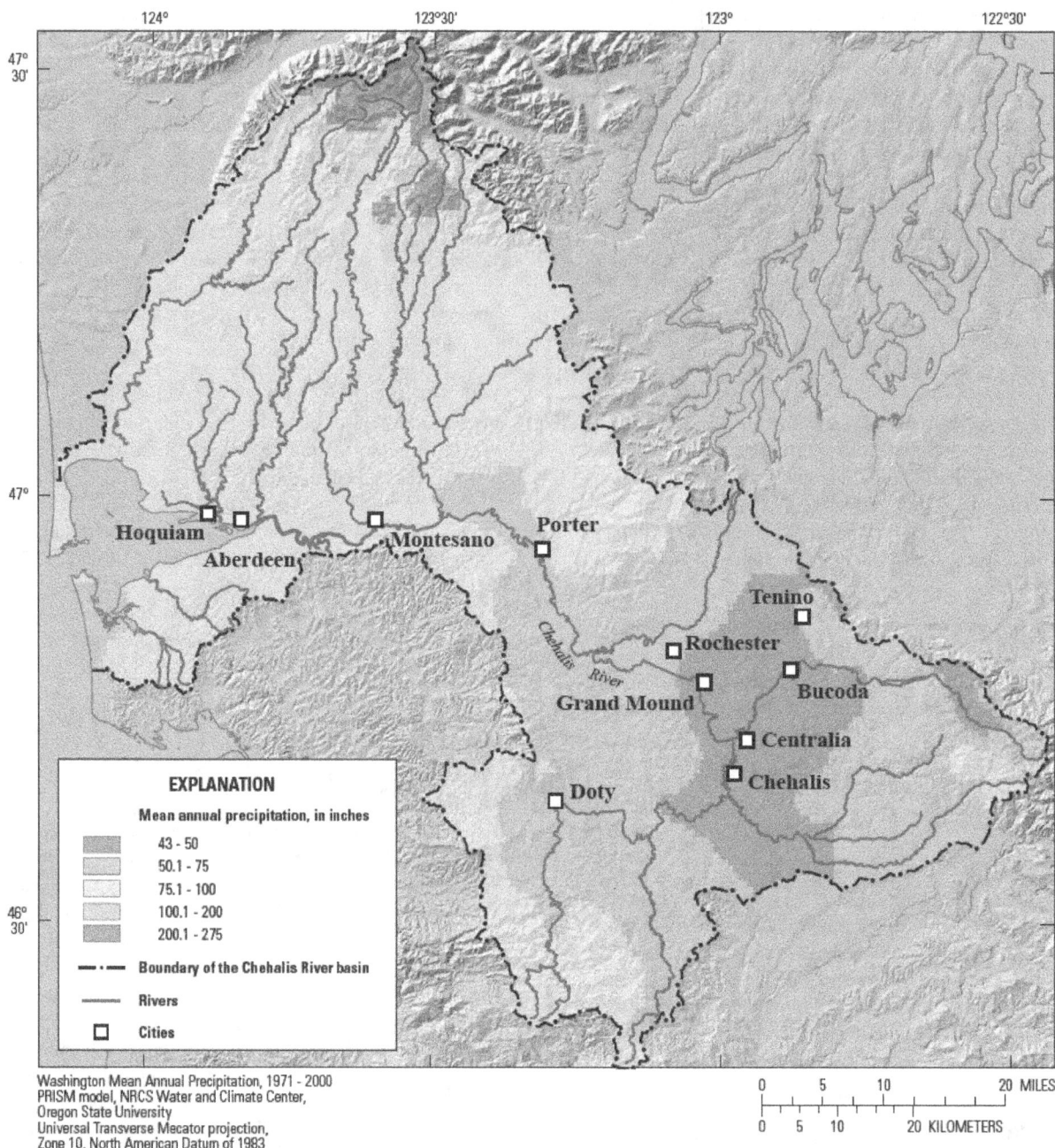

Figure 2.   Mean annual precipitation in the Chehalis River basin, southwestern Washington, 1971–2000.

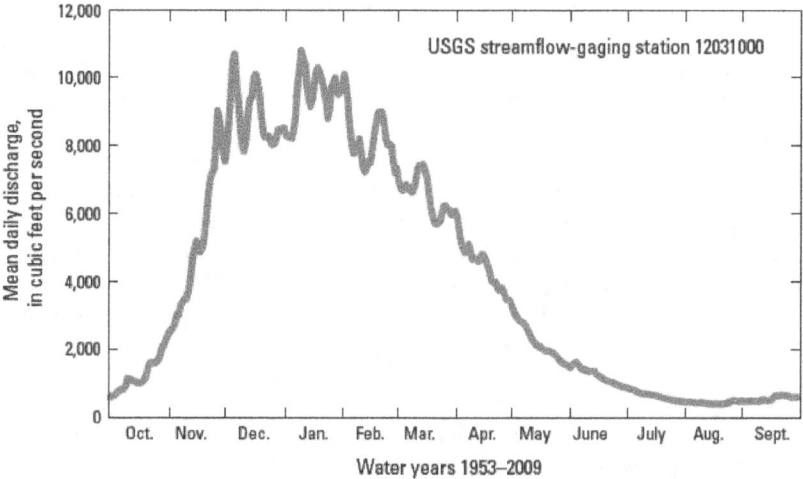

Figure 3.   Mean daily discharge at USGS streamflow-gaging station 12031000 on the Chehalis River at Porter, southwestern Washington, water years 1953–2009.

Previous Investigations

This investigation builds upon several previous studies of the hydrogeology of the Chehalis River basin, which varied in scope and purpose. The earliest studies published as Water Supply Bulletins by the State of Washington (Wallace and Molenaar, 1961; Weigle and Foxworthy, 1962; Eddy, 1966; Noble and Wallace, 1966) described regional hydrogeologic units and the occurrence and quality of groundwater. Several more recent studies built upon this work in the Scatter Creek and Black River area (Sinclair and Hirschey, 1992), northern Thurston County (Drost and others, 1998), and the Chehalis-Centralia area (Pitz and others, 2005). A regional characterization of surficial aquifers was also completed by Garrigues and others (1998). The hydrogeologic framework presented in this report generalizes the hydrogeologic characterizations presented in these previous investigations into a regional framework and considers more recent geologic mapping (Logan and others, 2009) and well-log information.

Methods of Investigation

Basic hydrologic and geologic data, including a field inventory of wells, measurement of water levels in wells, and extents of surficial geologic units were collected to delineate hydrogeologic unit maps, hydrogeologic cross-sections, and water-level maps. Several existing wells near the Chehalis River and its major tributaries were monitored at monthly intervals and some were instrumented with pressure transducers for comparison to stream stage measured at streamflow-gaging stations in order to determine the dynamics of groundwater/surface-water interaction during water year 2010. Streamflow was measured during a 3-day period at 41 locations along the Chehalis River and its tributaries to determine bulk streamflow gains and losses during summer low-flow conditions.

Well Inventory and Water-Level Measurements

A field inventory of 360 wells was completed between July and September 2009 to acquire information about the spatial distribution of the physical and hydraulic properties of the geologic units that comprise the groundwater-flow system of the Chehalis River basin (table 5, at back of report; Fasser and Julich, 2010). In addition, 12 wells from previous USGS studies were used to augment information provided by wells inventoried in 2009 for the construction of hydrogeologic cross-sections. The location of each well was surveyed using a Global Positioning System receiver with a horizontal accuracy of ±10 ft. The altitude of the land surface at each well location was interpolated from a U.S. Geological Survey (USGS) Digital Elevation Model with a 1/3 arc-second cell spacing and a vertical accuracy of ±8 ft . Well logs obtained from the Washington State Department of Ecology (WADOE) or well owners were linked with field inventory wells through WADOE unique well identification tags. Well-construction details including depth, screened interval, and lithologic units encountered were determined from drillers' well logs. Depth to water was recorded at most wells in accordance with standard USGS techniques (Drost, 2005) using a calibrated electric tape or graduated steel tape, both with accuracies of 0.01 ft. All well inventory data—including well location, construction, and water-level measurements–were entered into the USGS National Water Information System (NWIS) database.

A subset of 14 of the 360 inventoried wells located in the surficial aquifers near the Chehalis River were monitored at monthly intervals from November 2009 to October 2010 and entered into the NWIS database. Hourly water-level measurements were recorded at six wells within the monthly monitoring network by pressure transducers (In-Situ Level Troll 500). Wells within the monthly and hourly monitoring networks were selected to evaluate groundwater/surface-water interactions, and to document seasonal fluctuations in groundwater levels in surficial aquifers away from major withdrawal areas (table 5).

Hydrogeology

Surficial geology of the Chehalis River basin was simplified from existing surficial geologic maps including the 1:100,000 scale geologic maps of the Centralia quadrangle (Schasse, 1987), the Chehalis River and Westport quadrangles (Logan, 1987), and the Shelton quadrangle (Logan, 2003). These maps were compiled into a digital map database by the Washington State Division of Geology and Earth Resources (2005). Recent 1:24,000 geologic mapping by Logan and others (2009) provided additional details of the Maytown Quadrangle in southern Thurston County. Modifications to the extent of existing mapped units were made in some areas based on lithologic information from drillers' lithostratigraphic logs obtained during this study.

Seepage Investigation

Forty-one discharge measurements were made during low-flow season to capture baseflow conditions over a 3-day period in August 2010 along the Chehalis, Skookumchuck, and Black Rivers, Scatter Creek, and select smaller tributaries (U.S. Geological Survey, 2011). The net volume of water exchanged between the surface-water and groundwater systems over the stream length between discharge measurement sites—termed the seepage reach—was calculated as the increase or decrease in streamflow that is not accounted for by tributary inflows or diversions between the two discharge measurement sites. Calculated streamflow gains and losses are expressed as the net volume of water exchanged over the seepage reach in cubic feet per second, and the net volume of water exchanged normalized by the seepage reach length in cubic feet per second per mile.

Most discharge measurements used to calculate groundwater/surface-water exchanges were measured on the same day, but some measurements were made up to 2 days apart, introducing uncertainty of the same magnitude as measurement error into the groundwater/surface-water exchange estimates. Mean daily discharge varied by 19 ft^3/s, or 8 percent of mean daily streamflow (237 ft^3/s) during the seepage investigation, at the Chehalis River near Grand Mound (USGS gage 12027500), and by 18 ft^3/s, or 4 percent of mean daily discharge (449 ft^3/s), during the seepage investigation at the Chehalis River at Porter (USGS gage 12031000) (fig. 4).

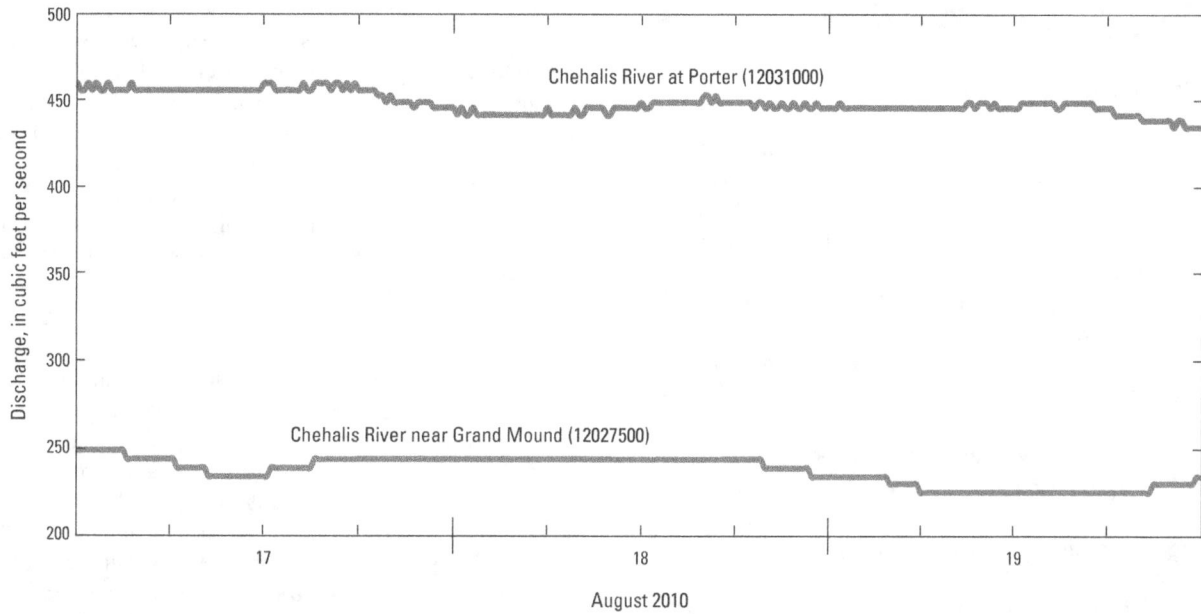

Figure 4.   Discharge for two U.S. Geological Survey streamflow-gaging stations, Chehalis River basin, southwestern Washington, August 17–19, 2010.

Discharge was measured using the velocity-area method using standardized USGS techniques with a Price AA meter (Rantz, 1982) or an acoustic Doppler current profiler (ADCP; Oberg and others, 2005). Measurement sites inaccessible by vehicle were measured by a two-person team using a kayak-mounted ADCP to traverse the river, allowing for an increased density of seepage measurements in this study. Each discharge measurement was assigned an accuracy rating of "good," indicating measurements are within 5 percent margin of error; "fair," indicating measurements are within 8 percent margin of error; and "poor," indicating that measurements have an error of 8 percent or more (Sauer and Meyer, 1992). The measurement error associated with the upstream, downstream, tributary, and diversion streamflow measurements used to calculate a single seepage gain or loss was propagated using the following formula for each seepage reach (Wheeler and Eddy-Miller, 2005):

$$s = \sqrt{(\pm a)^2 + (\pm b)^2 + \dots (\pm n)^2},\tag{1}$$

where

s is the prorogated error of the individual of the $a, b, \dots, n$ discharge measurements associated with seepage gain or loss calculation.

Many small surface-water diversions and returns of up to 4.2 and 5.7 ft^3/s, respectively, occur within the Chehalis River and its tributaries, but their quantification was outside the scope of this study. These unmeasured diversions and returns are most active during the summer irrigation season and introduce some error into the calculated streamflow gains and losses.

Hydrogeologic Framework

Geologic Setting

The Chehalis River basin occupies the southern margin of the Puget Lowland, a glacially modified structural basin, and extends into the foothills of the Cascade Mountains, and the Willipa Hills and Olympic Mountains of the Washington Coast Ranges. Tectonic and glacial processes during the Tertiary and Quaternary periods have shaped the present topography, stratigraphy, and structure of the Chehalis River basin (fig. 5).

Tertiary to present convergence of the North American continental plate and the Farrallon and Juan de Fuca oceanic plates has resulted in the formation and maintenance of a subduction zone and volcanic arc along the western margin of the North American plate (Wells and others, 1984). Marine sediments and several exotic terranes were scraped off the leading edge of the subducting oceanic plates prograding the margin of North America westward. The Tertiary volcanic, as well as the marine and non-marine sedimentary, rocks deposited along this margin were uplifted to their present position as a result of this convergence resulting in dominant southeast-northwest trending faults and folds (Snavely and others, 1958). The Juan de Fuca Plate is currently converging at a rate of 1–1.5 in/yr obliquely towards the North American Plate (Heaton and Kanamori, 1984).

During the Pleistocene Epoch of the Quaternary Period, the Puget Lobe of the Cordilleran Ice Sheet advanced into the northwestern portion of the Chehalis River basin at least twice (Bretz, 1913; Lea, 1984), most recently about 15,000 years ago (15 Kya) during the Vashon Stade (glacial readvance) of the Fraser Glaciation. The pre-existing bedrock topography surrounding the Puget Lowland including the Olympic and Cascade Mountains determined both the extent of the Puget Lobe and its meltwater drainage. After the advance of the Puget Lobe southward of the Strait of Juan de Fuca, large proglacial lakes formed in the Puget Lowland, eventually draining through an outlet at the southern terminus of the Puget Lowland through the modern Chehalis River valley and depositing thick sequences of advance glacial outwash. The Puget Lobe during Vashon time reached its maximum extent ca. 13.5 Kya, forming terminal moraines north of Rochester and depositing till farther to the north. During deglaciation, meltwater was routed through a series of spillways and valleys (Logan and others, 2009), dissecting the Vashon terminal moraine and depositing recessional glacial outwash within the Chehalis River valley and the valleys of several of its tributaries, primarily Scatter Creek and the Skookumchuck, Black, and Satsop Rivers. Coarse-grained recessional outwash was deposited as far south as Centralia, damming the Chehalis River valley and its tributaries upstream of this point to form glacial Lake Chehalis (Bretz, 1913) where fine-grained lacustrine sediments were deposited. The Puget Lobe occupied the Chehalis River basin at least once before during the Fraser Glaciation, but extended 1–7 mi beyond its Vashon extent more than 125 Kya, depositing a drift sequence termed the Penultimate drift by Lea (1984) as far south as Centralia. Unlike Vashon-age drift, which crops out widely in the northern Chehalis River basin, the Penultimate drift has limited surface exposure and its original topographic form is less well-preserved because it has been eroded or covered by Vashon-age drift.

Alpine glaciers originating in the Olympic and Cascade Mountains also advanced several times during the Pleistocene into the Chehalis River basin, depositing several sequences of the drift in headwaters and valleys. The Logan Hill drift was deposited as an outwash plain or valley train unconformably on Tertiary rocks by alpine glaciers originating from the Cascades and consists of Cascade-derived volcanic and sedimentary rocks of Tertiary age (Snavely and others, 1958) prior to the development of the modern volcanic cone of Mt. Rainier (Easterbrook, 1986). No radiometric dates or paleomagnetic data exist, but Crandell (1963) correlates the Logan Hill drift with the Orting Drift deposited by the Puget Lobe of the Cordilleran ice sheet during the Early Pleistocene. Most of the Logan Hill drift consists of sand and gravel outwash, but isolated layers of till also exist (Crandell and Miller, 1974). Subsequent erosion has reduced the extent of the Logan Hill drift to the bedrock uplands, and the upper 24–74 ft has been intensely weathered to clay (Snavely and others, 1958). Outwash from two additional advances of Cascade alpine glaciers was deposited during the Middle Pleistocene as the Hayden Creek and Wingate Hill drifts, although the glaciers terminated outside of the modern Chehalis River basin (Crandell and Miller, 1974). The Wingate Hill and Hayden Creek drifts are much less areally extensive than the Logan Hill drift and are confined to terraces bordering Newaukum and Chehalis River valleys.

Alpine glaciers have advanced from the Olympic Mountains at least four times during the Pleistocene (Logan, 2003 and references therein), depositing across the northwestern portion of the Chehalis River basin till and outwash composed of marine sedimentary and volcanic rocks derived from the interior of the Olympic Mountains. Logan (2003) correlated locally named drift sequences of previous workers across the southern Olympic Mountains, grouping them into drift from four Pleistocene glacial advances: two during Wisconsinan time, and two prior to Wisconsinan time. Although the relative chronology of Olympic Mountain alpine glacial advances has been studied considerably, the absolute chronology remains relatively unknown except for alpine glacial advances farther to the north in the Hoh River and Queets River basins (Thackray, 1996).

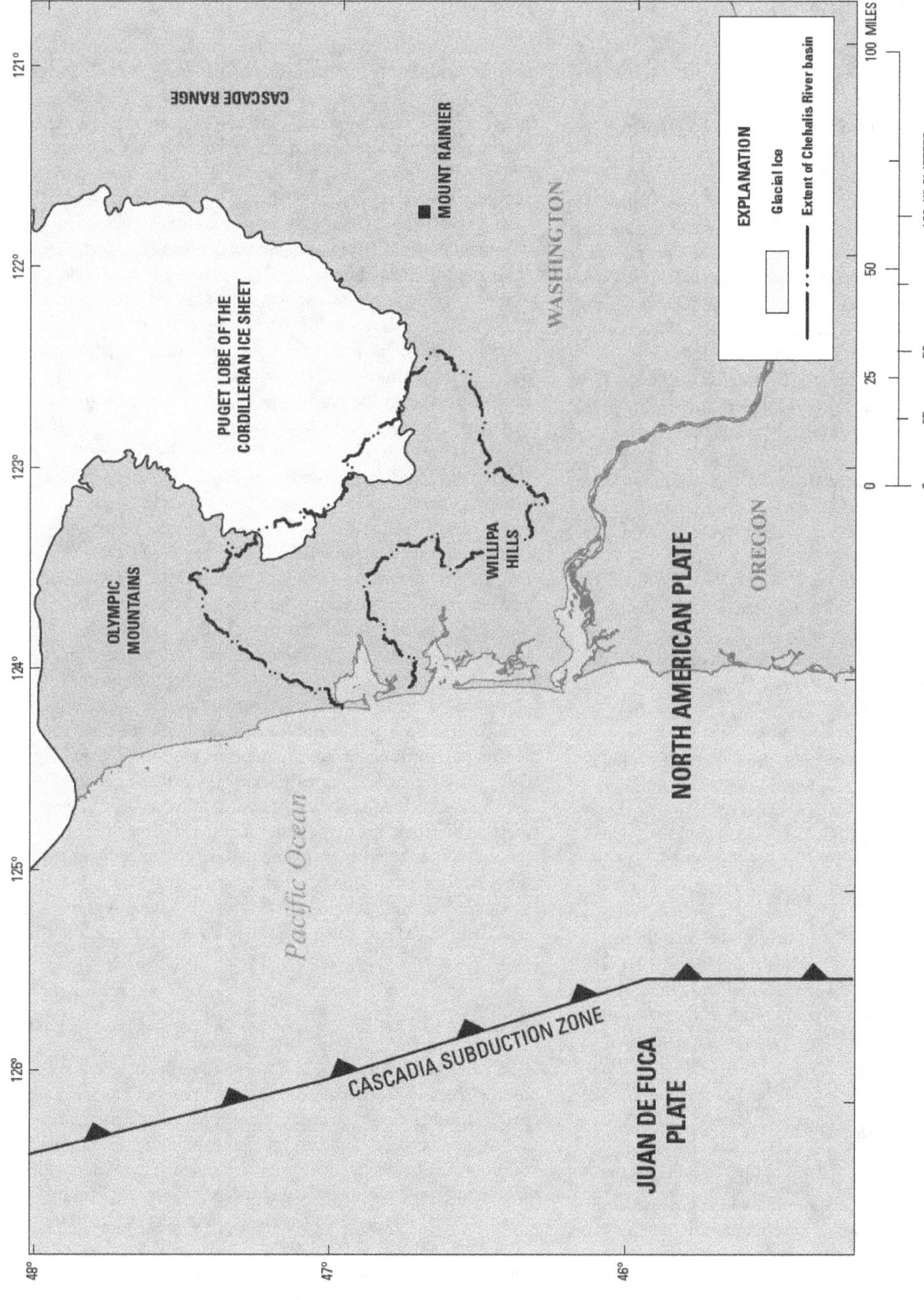

Figure 5.   Extent of Puget Lobe of the Cordilleran Ice Sheet and location of Cascadia Subduction Zone in western Washington and Oregon.

Hydrogeologic Setting

Geologic materials capable of yielding water in significant quantities to wells or springs are classified as aquifers, whereas lower permeability geologic materials that limit the movement of groundwater are classified as confining units. The presence or absence of confining units in stratigraphic relation to aquifers determines whether an aquifer is confined or unconfined. Unconfined, or "water-table," aquifers occur where the saturated zone of the aquifer can equilibrate to atmospheric pressure and the upper surface of the aquifer (water table) rises and declines in response to groundwater recharge and discharge. Confined, or "artesian," aquifers are overlain by a lower permeability confining unit that prevents water from equilibrating with atmospheric pressure. The water level in a properly sealed well drilled into a confined aquifer will rise above the top of the aquifer to the level of the potentiometric surface, which is the hydraulic head of the aquifer. The potentiometric surface, like the water table in an unconfined aquifer, rises and declines in response to recharge and discharge of the aquifer. Both unconfined and confined aquifers provide significant sources of groundwater in the Chehalis River basin.

The primary aquifers in the Chehalis River basin are comprised of Pleistocene glacial outwash and Holocene alluvium deposited along the valleys of the Chehalis River and its major tributaries. Intervening deposits of Pleistocene glacial till and fine-grained inter-glacial sediments act as semi-confining to confining layers. The upper portion of the alpine glacial outwash within the uplands of the Chehalis River basin has been weathered to clay, forming a confining unit above the alpine glacial outwash. Although these units act regionally as aquifers and confining layers, a large amount of heterogeneity exists within the Quaternary glacial and non-glacial sediments, resulting in localized areas of high permeability within confining units, or low hydraulic conductivity within aquifers. The areal extent of the hydrogeologic units is determined by the Quaternary topography of the Chehalis River basin and the extent of Pleistocene glaciations of both continental and alpine origins. The Tertiary bedrock, consisting of marine and continental sediments, as well as basalts, have relatively low hydraulic conductivities, forming a low permeability basal unit.

Hydrogeologic Units

Hydrogeologic units were differentiated on the basis of the lithologic properties, hydrologic characteristics, and relative stratigraphic position of rocks and sediments in the Chehalis River basin. Five hydrogeologic units above a low-permeability basal bedrock unit were identified in the Chehalis River basin (pl. 1 and table 1) and are described below.

A Aquifer

The A aquifer extends throughout the major river valleys and lowland prairies of the Chehalis River and its tributaries and comprises the most areally extensive surficial aquifer. This aquifer interacts readily with surficial water features, in general receiving recharge from rivers during the winter when river stages are high, and discharging to rivers in the summer when river stages are low. This unit contains silt, sand, gravel, and coarser alluvial sediments of glacial and non-glacial origin. The youngest sediments in the A aquifer are coarse-grained channel and fine-grained overbank deposits of modern rivers, which are distributed across the floodplain of the Chehalis River and its tributaries. Modern alluvium overlies and is in direct hydrologic connection with glacial outwash from alpine glaciers, outwash from the Puget lobe of the Cordilleran Ice Sheet, and glacio-lacustrine sediment deposited in pro-glacial lakes.

Although significant heterogeneity exists within the A aquifer, including the presence of local confining layers, several generalizations about the character and hydraulic properties of sediments within this aquifer can be made. First, within the Chehalis River valley north of Centralia and the prairies surrounding Scatter Creek and the Black River, advance and recessional glacial outwash deposited during the Vashon stade (glacial readvance) of the Cordilleran Ice Sheet comprise a distinct coarse, well-sorted portion of the A aquifer. Second, south of Centralia, the aquifer is comprised of fine-grained sediment and poorly sorted and weathered alpine glacial outwash of Hayden Creek age (140 Kya), forming terraces in the Newaukum River and Chehalis River valleys. Pre-Vashon outwash units in the D undifferentiated deposits are in hydrologic connection with the A aquifer and are difficult to distinguish in well logs; therefore, the D undifferentiated deposits also are included in the A aquifer. Holocene beach deposits also are included in the A aquifer. Bedrock and pre-Vashon till layers within the C undifferentiated deposits form the basal confining unit of this aquifer.

B Confining Unit

The B confining unit is distributed in the northern part of the Chehalis River basin and is comprised of unsorted and unstratified clay- to boulder-sized particles. Irregularly distributed layers of sand and gravel containing small amounts of groundwater occur locally. Although some local deposits of sand and gravel produce local aquifers, the B unit primarily is comprised of fine-grained sediments, and acts as a confining unit. This unit was deposited during the last glacial advance at the southern margin of the Puget Lobe of the Cordilleran Ice Sheet. The Black Hills separates two distinct areas of till and end moraines that comprise the B confining unit.

Table 1. Hydrogeologic units in the Chehalis River basin, southwestern Washington.

[Hydrogeologic units defined in this study are delineated in plate 1. Hydrogeologic units of previous studies are defined in cited references. **Abbreviations:** –, not differentiated]

Period	Epoch	Hydrogeologic units defined in this study	Range of thickness [estimated average thickness] (feet)	Number of wells open to unit	Hydrogeologic units of previous studies				
					Drost (1998)	Pitz and others (2005)	Weigle and Foxworthy (1962)	Eddy (1966)	Noble and Wallace (1966)
Quaternary	Holocene to Pleistocene	A	4–150 [20]	100	Qvr	Qa, Qgo(g), Qapo(h)	Qal, Qt, Qo, Qnt, Qlc	Qal, Qb, Qtv, Qto	Qal, Qvr, Qvrl
	Pleistocene	B	5–52 [21]	0	Qvt, Qvrm	–	–	–	Qvm, Qvt
		C	25–42 [36]	18	Qva	–	–	–	Qva
		D	16–203 [91]	41	–	Qapo(lh), Qal	Qlh	QTu	Qlh
		E	18–15 [100]	42	Qf,Qc,TQu	Qago(g)	–	–	Qss1, Qss2, Qk, Qpu
Tertiary	Eocene to Pliocene	BDRK	Not applicable	149	Tb	Mc(w), Tb(bslt), Tbu, Qal	Tu, Tcr	Tu	Ts, Tv

C Aquifer

The C aquifer is comprised primarily of well-sorted sand, gravel, and cobble-sized sediment that was deposited as advance glacial outwash during the Vashon Stade of the Fraser Glaciation. Pre-Vashon age outwash in hydrologic connection also is included in the C aquifer. The C aquifer is confined by the B confining unit in the northern part of the Chehalis River basin.

D Undifferentiated Aquifers and Confining Units

The D unit is comprised of Pre-Vashon tills and outwash sequences deposited in the northern Chehalis River basin as far south as Centralia. Groundwater within the D unit occurs under confined conditions within coarse-grained outwash sequences that are separated from stratigraphically higher aquifers A and C by thin till layers. Multiple aquifers and confining units within the D unit may exist where they have not been eroded and have not been differentiated because they are not laterally continuous. Outwash of this unit is included within the A and C aquifers when in direct hydrologic connection with those aquifers.

E Aquifer

Alpine glacial outwash emanating from the Cascade and Olympic Mountains comprises the E aquifer on the bedrock uplands of the Chehalis River basin. Several episodes of alpine glaciation have been documented since the early Pleistocene, and their deposits consist of Cascade and Olympic-derived Tertiary volcanic and sedimentary rocks, including siltstones, sandstones, and conglomerates. The top portions of these deposits have been extensively weathered into clay, confining groundwater where this unit is saturated.

BDRK Low Permeability Basal Bedrock Unit

The Tertiary bedrock (BDRK) forms the basal confining unit of the groundwater-flow system and is relatively impermeable in relation to the unconsolidated sediments stratigraphically above it. Marine and non-marine siltstones, sandstones, and conglomerates locally containing coal beds comprise the Tertiary bedrock unit along with intrusive and extrusive volcanic rocks. Bedrock locally yields water sufficient for domestic use through fracture flow. In the southern part of the Chehalis River basin, appreciable quantities of groundwater are contained within sandstone interbeds of the inferred non-marine Miocene Wilkes Formation, which are confined by overlying clay layers (Pitz and others, 2005). This groundwater may be under artesian pressure where it is present.

Hydraulic Conductivity

Hydraulic conductivity (the ability of a material to transmit water) depends on the physical properties of the material, including the size, sorting, shape, and compaction of particles. The glacial and alluvial sediments that form the primary aquifers and confining layers within the Chehalis River basin have a large degree of heterogeneity of these physical properties, creating variations of hydraulic conductivity ranging over several orders of magnitude. The magnitude and distribution of horizontal hydraulic conductivities were estimated for wells where specific capacity data (water-level drawdown over time after pumping a well for a specified period of time) were available from drillers' logs. Wells with less than 2 hours of pumping during a specific-capacity test were excluded from analysis. Hydraulic conductivity was estimated from specific capacity tests using one of two sets of equations, depending on whether the well was completed as an open-ended casing, or a screened/perforated interval.

An equation developed by Bear (1979) was modified for spherical flow to an open-ended well:

$$K_h = \frac{Q}{4\pi s r}, \tag{2}$$

where

K_h is horizontal hydraulic conductivity, in feet per day;

Q is the pumping rate, or discharge, of the well, in cubic feet per day;

s is the drawdown in the well, in feet; and

r is the radius of the well, in feet.

This equation includes the assumption that horizontal and vertical hydraulic conductivities are equal. This assumption is likely violated because horizontal hydraulic conductivity is likely greater than vertical hydraulic conductivity in aquifers in the Chehalis River basin. Therefore, this equation may underestimate K_h values.

For wells with a perforated interval, K_h was calculated as the quotient of the transmissivity (T), in days, and the aquifer thickness approximated by the length of the perforated interval, in feet. The modified Theis equation (Ferris and others, 1962) was used to calculate transmissivity for perforated interval wells:

$$s = \frac{Q}{4\pi s T} \ln \frac{2.25 T t}{r^2 S}, \tag{3}$$

where

s is the drawdown in the well, in feet;

Q is the pumping rate, or discharge, of the well, in cubic feet per day;

T is the transmissivity of the hydrogeologic unit, in days;

t is the length of time the well was pumped, in days;

r is the radius of the well, in feet;

S is the storage coefficient, a dimensionless number, assumed to be 0.0001 for confining layers; and 0.1 for aquifers.

This equation, which assumes horizontal flow, does not account for flow from above or below the perforated interval, and may result in an overestimation of hydraulic conductivity.

Horizontal hydraulic conductivities of the hydrogeologic units of the Chehalis River basin vary over several orders of magnitude (table 2). Values of K_h were greatest for the unconsolidated alluvial and Puget Lobe glacial outwash aquifers (A, C, and D), and were smaller for the E aquifer, which represents older Cascade and Olympic-mountain derived glacial outwash that has been deeply weathered to clay locally. Specific capacity test data were not available for the B hydrogeologic unit, but it is assumed to have a low hydraulic conductivity based on its primarily unsorted texture and lithologic composition. The bedrock low permeability basal unit (BDRK) had the lowest estimated hydraulic conductivities with fracture flow as an important component.

Table 2. Estimates of hydraulic conductivity in the Chehalis River basin, southwestern Washington.

[Hydrogeologic units are delineated in plate 1. ft/d, feet per day; –, not determined]

Hydrogeologic unit	Number of wells	Horizontal hydraulic conductivity (ft/d)		
		Minimum	Maximum	Median
Estimated from specific-capacity data for perforated interval wells				
A aquifer	11	2.2	399.9	66.3
B confining unit	0	–	–	–
C aquifer	2	1.1	88.2	44.6
D undifferentiated	2	8.9	490.6	249.8
E aquifer	2	0.2	1.8	1
BDRK basal confining unit	15	0.001	1.5	0.1
Estimated from specific-capacity data for open ended wells				
A aquifer	3	122.5	229.8	122.5
B confining unit	0	–	–	–
C aquifer	0	–	–	–
D undifferentiated	2	153.2	294.1	223.7
E aquifer	1	–	–	36.8
BDRK basal confining unit	5	0.8	76.6	9.7

Groundwater Movement and Fluctuation

A regional water-level contour map for the surficial aquifers, including the A and D units, was drawn from water levels measured between July and September 2009 to estimate general directions of horizontal groundwater flow (fig. 6). Wells completed in the D aquifer were included within the delineation of this map because pre-Vashon age glacial till layers confining groundwater in the D unit are discontinuous and sometimes absent, putting A and D in close hydrologic connection. Groundwater flows perpendicular to water-level contours (water-table altitude in an unconfined aquifer) from areas of higher to lower water levels. In general, horizontal groundwater flow follows the contours of the surface-water drainage of the Chehalis River and its tributaries, flowing from the headwaters towards Grays Harbor. Hydraulic gradients are relatively steep in the tributary valleys such as the Newaukum Valley (about 23 ft/mi) and flatter in the alluvial valley of the central Chehalis River (about 6 ft/mi). The lowest groundwater levels in the surficial aquifers occur in the lower Chehalis

River basin near the confluence of the Satsop and Chehalis Rivers. The tides of Grays Harbor may account for elevated groundwater levels in the lower Chehalis River basin.

Groundwater levels in the Chehalis River basin fluctuate seasonally and over shorter timescales in response to changes in aquifer recharge and discharge driven by storm events, river stage, and tidal fluctuations. In general, water levels rise during autumn and winter when precipitation is highest and water use is lowest; conversely, water levels decline during the spring and summer when precipitation is lower and water use increases. Water levels typically were lowest in late summer to early fall. Water levels measured in 14 wells completed in the surficial aquifer fluctuated by as little as 4.8 ft to as much as 16.8 ft during water year 2010. The largest groundwater fluctuation occurred in well 15N/03W-03A02P3 (fig. 7A). This well was the farthest from a river of the monitored wells and was completed within the D aquifer confined beneath a till of pre-Vashon Age. Rivers and other surface-water features may attenuate water-level fluctuations within riparian wells with direct connection to surface-water systems because they can provide aquifer recharge when the aquifer levels are low and receive aquifer discharge when aquifer levels are high.

Three sites within the monthly monitoring network were constructed with three nested wells open at different levels within the D unit across multiple low-permeability layers. This allowed quantification of vertical gradients at these three locations. At the first site containing wells 15N/03W-03A02P1, 15N/03W-03A02P2, and 15N/03W-03A02P3, the water level measured at the deepest well 15N/03W-03A02P1 (125 ft deep) was consistently lower (0.46–0.99 ft; median: 0.74 ft) than the water level measured at the shallowest piezometer 15N/03W-03A02P3 (52 ft deep) indicating downward movement of water throughout the year (fig. 7A). At the second site containing wells 15N/03W-10D02P1, 15N/03W-10D02P2, and 15N/03W-10D02P3, water levels measured at monthy intervals within the deepest well (15N/03W-10D02P1; 83 ft deep) also were consistently lower than the shallowest well (15N/03W-10D02P3; 35 ft deep) by 0.05–0.37 ft (median: 0.21 ft; fig. 7B). Water-level gradients determined at the third site containing wells 16N/03W-33P01P1, 16N/03W-33P01P2, and 16N/03W-33P01P3 agreed with the first two sites with water levels at the deepest well 16N/03W-33P01P1 (93 ft deep) consistently lower than water levels within the shallowest well (16N/03W-33P01P3) (46 ft deep) by 0.23–2.4 ft (median: 0.29 ft; fig. 7C). These water-level differentials show that vertical flow does occur through confining to semi-confining units within the unconsolidated sediments of the D Unit.

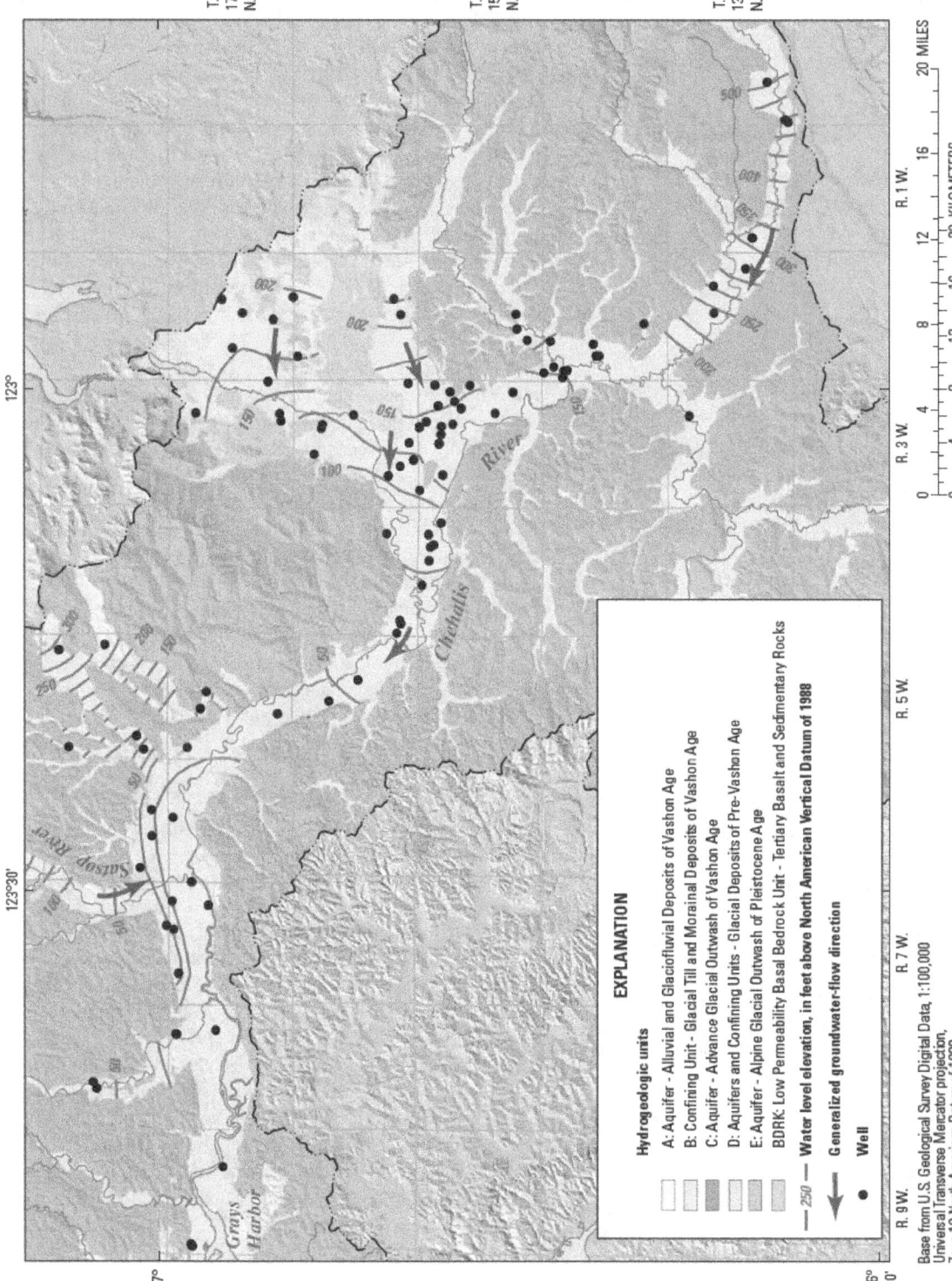

Figure 6. Water-table altitudes and inferred direction of groundwater flow in the surficial aquifers, Chehalis River basin, southwestern Washington, August–September 2009.

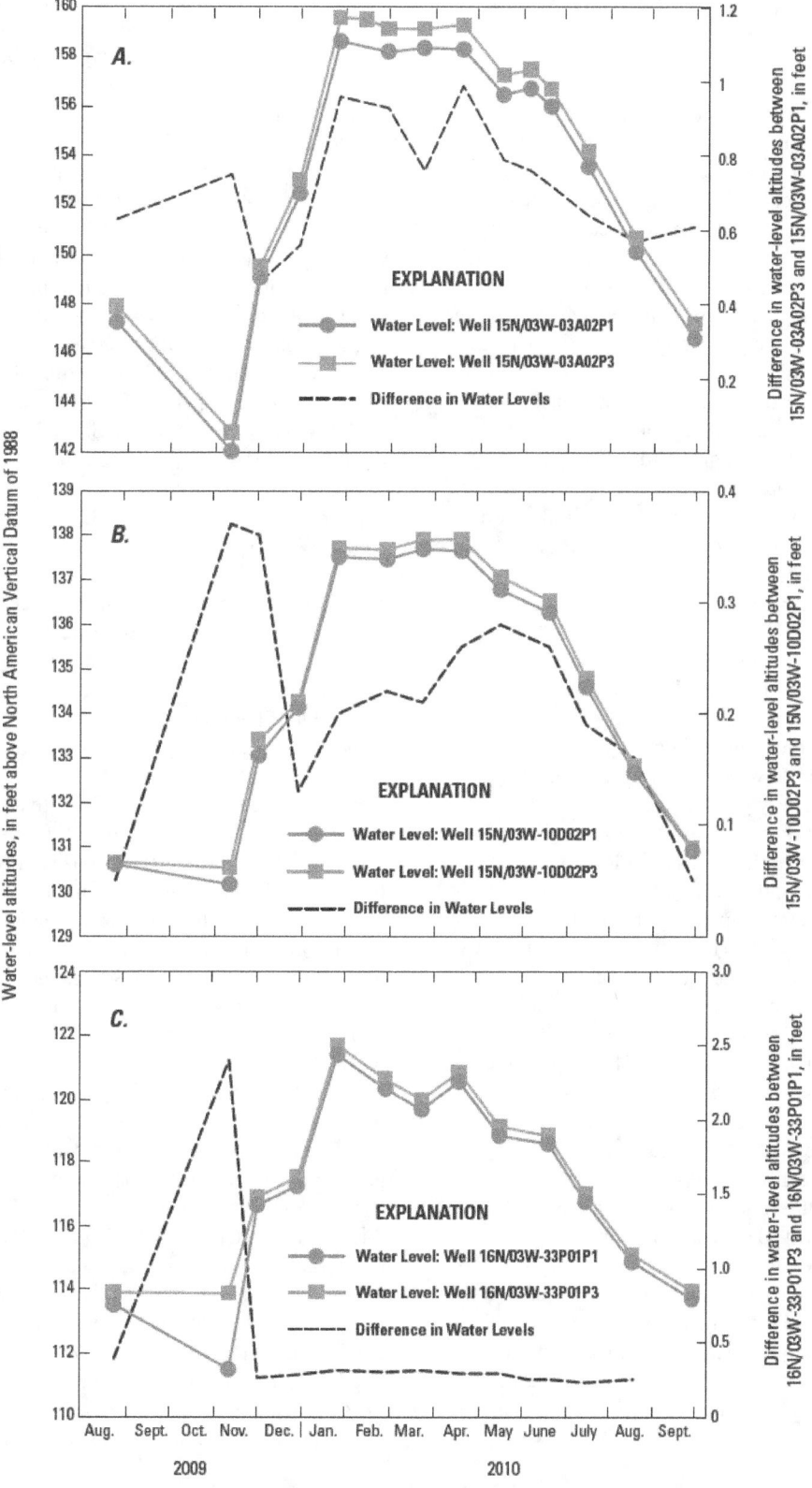

Figure 7.   Water-level altitudes for nested wells (A) 15N/03W-03A02P1 and 15N/03W-03A02P3, (B) 15N/03W-10D02P1 and 15N/03W-10D02P3, and (C) 16N/03W-33P01P1 and 16N/03W-33P01P3, Chehalis River basin, Washington, August 2009–October 2010.

Groundwater/Surface-Water Interactions

Surface water in rivers and lakes can readily interact with groundwater in aquifers, resulting in the exchange of appreciable quantities of water and solutes. This exchange, or seepage, provides recharge of aquifers and maintenance of streamflows, and has the potential to affect the quality of groundwater and surface-water bodies. When the water table in an unconfined aquifer is higher than the river stage, an upward hydraulic gradient drives the movement of water from the aquifer to the river, resulting in a gaining stream; conversely, a downward hydraulic gradient exists when the river elevation is higher than the water table, causing the water to seep from river to the underlying aquifer.

The Chehalis River is a rainfall-runoff dominated system in which the seasonal variation in streamflow is greater than 100 percent of the mean annual streamflow of 4,049 ft^3/s (fig. 3). The relative contribution of groundwater to streamflow varies significantly between seasons. Pitz and Sinclair (1999) calculated baseflow, a measure of groundwater contribution to streamflow, at the Chehalis River at Porter (USGS site No. 12031000). Pitz and Sinclair (1999) determined that baseflow varies from only 45 percent of total streamflow in November to as much as 89 percent of total streamflow in July, when streamflows are near their annual minimum.

A series of streamflow measurements, termed a seepage run, were made to identify gaining and losing reaches in the Chehalis River during summer baseflow conditions in August 2010 when streamflow was at its lowest (fig. 3) and the contribution of groundwater as a fraction of streamflow was likely at its annual maximum. Synoptic discharge measurements were made at 41 locations and streamflow gains and losses were calculated for 28 reaches of the Chehalis River from its confluence with the Newaukum River to Oakville, Skookumchuck River, Scatter Creek, and Black River from August 17–19, 2010. The seepage reaches in this study are within the extent of a 2007 USGS seepage investigation (Ely and others, 2008) and three previous studies (Sinclair and Hirschey, 1992; Pitz and others, 2005; Ely and others, 2008). Additional streamflow gaging locations were incorporated within this study, resulting in a more detailed seepage evaluation, with 18 seepage reaches along the central Chehalis River, which ranged in length from 0.8 to 8.4 mi. Additionally, wells within the surficial aquifer were measured at hourly and monthly intervals from November 2009 to August 2010 to characterize the magnitude and timing of seasonal fluctuations in groundwater levels in relation to exchanges with surface-water bodies and recharge from precipitation.

The Chehalis River gained 56.8 ± 23.7 ft^3/s over 32.8 river miles [1.7 (ft^3/s)/mi] from the upstream to downstream extent of the seepage run, but most of the incremental gains and losses were within compounded measurement error (table 3; fig. 8). Only one seepage reach (fig. 8; reach F) showed a net gain of 29.0 ± 25.5 ft^3/s [17.1 (ft^3/s)/mi] , and only two seepage reaches showed net losses of -48.3 ± 24.5 ft^3/s [-48.3 (ft^3/s)/mi] (reach O) and -23.0 ± 22.5 ft3/s (-17.7 (ft^3/s)/ mi], (reach K), which were larger in magnitude than the associated compounded measurement errors. The small number of significant gains and losses relative to compounded measurement error indicates diffuse groundwater inflow between reaches D and G.

Ely and others (2008) report that the largest river-aquifer exchanges on the Chehalis River observed in September 2007 occured downstream of Grand Mound where the Chehalis River enters a broad prairie of low relief. Mean daily streamflow was approximately 120 ft^3/s lower during the September 2007 seepage run (329 ft^3/s at Chehalis River at Porter, Washington: 12031000) than during the August 2010 seepage run (449 ft^3/s at Chehalis River at Porter, Washington: 12031000). During the August 2010 seepage run, the Chehalis River remained near neutral upstream of river mile 58.8 before gaining 29 ft3/s within seepage reach F (fig. 9; RM 57.1–58.8). Seepage reach F had the highest gradient (0.0012 ft/ft) within the extent of the seepage run (mean gradient: 0.0005 ft/ft) potentially gaining streamflow by intersecting the water table of the underlying unconfined aquifer. A longer seepage reach measured in September 2007 that encompassed seepage reaches E, F, G, and H gained 76.9 ft^3/s suggesting that this section of the Chehalis River consistently gains streamflow from the underlying aquifer during low-flow conditions (fig. 9). The Chehalis River exchanged comparable amounts of water with the underlying aquifer in August 2010 as it did in September 2007 (fig. 9). Several notable exceptions exist including losses within seepage reaches I and R measured in 2007 that were not measured in 2010 and a loss in seepage reach O measured in 2010 that was not observed in 2007. This suggests temporal variability in river-aquifer exchanges within the centreal Chehalis River basin during low-flow conditions.

The central Chehalis River lacks common geologic and geomorphic controls that focus large exchange of groundwater and surface water. Changes in unconsolidated aquifer thickness, contact of lithologic units that differ markedly in hydraulic conductivity, and channel forms that increase hydraulic gradient between a river and shallow groundwater promote groundwater/surface-water exchanges (Konrad, 2006). Within the extent of the seepage run the depth of the alluvial sediment remained relatively constant (pl. 1, section B–B') precluding seepage gains associated with thinned alluvial sediments at the outlets of structural bedrock basins.

Table 3. Evaluation of gains and losses, with error analysis measurements for the seepage investigation, central Chehalis River Basin, southwestern Washington, August 2010.

[Associated measurement error for net gain or loss was calculated using the propagation of error formula (Wheeler and Eddy-Miller, 2005) where s is the error propagated from all estimated individual errors and a, b, ..., n are estimated errors for the discharge measurement at each site: $s = \sqrt{(\pm a)\,x^2 + (\pm b)\,x^2 + ...(\pm n)\,x^2}$). **Seepage reach**: Seepage reach delineated in figure 9. **Remarks**: Near neutral— Difference in measured discharge is less than associated measurement error. ft^3/s, cubic feet per second; (ft^3/s)/mi, cubic feet per second per mile; mi, mile; –, no data]

Measurement sites and No.	Measured streamflow (ft^3/s) Mainstem	Tributary	Assumed measurement error (percent)	Date	Net gain or loss (ft^3/s)	Associated measurement error (ft^3/s) +/-	Chehalis river mile (mi)	Seepage reach	Seepage reach length (mi)	Net gain or loss per river length [(ft^3/s)/mi]	Remarks
Chehalis River below Newaukum River, near Chehalis (12025025)	157.0	–	5	8/17/2010	–	–	75.1	–	–	–	–
Salzer Creek at Centralia Alpha Road, near Centralia (12025310)	–	0.0	11	8/19/2010	–	–	–	–	–	–	–
Coal Creek at National Avenue, at Chehalis (12025360)	–	0.3	11	8/17/2010	–	–	–	–	–	–	–
Skookumchuck River at mouth, near Centralia (12026620)	–	73.4	8	8/17/2010	–	–	–	–	–	–	–
Chehalis River at Fort Borst Park, near Centralia (12026625)	245.0	–	5	8/17/2010	14.3	15.7	66.7	A	8.4	1.7	Near neutral
Scammon Creek at Cooks Hill Road, near Centralia (12026670)	–	0.4	8	8/17/2010	–	–	–	–	–	–	–
Chehalis River at Galvin (12026700)	229.0	–	8	8/18/2010	-16.4	22.0	64.2	B	2.5	[1]-6.5	Near neutral
Lincoln Creek near Galvin (12027220)	–	1.9	8	8/17/2010	–	–	–	–	–	–	–
Chehalis River near Galvin (12027290)	229.0	–	5	8/17/2010	-1.9	21.6	61.4	C	2.8	[1]-0.7	Near neutral
Chehalis River near Grand Mound (12027500)	249.0	–	8	8/17/2010	20.0	23.0	59.9	D	1.5	13.3	Near neutral
Chehalis River above Prairie Creek, near Grand Mound (12027505)	249.0	–	5	8/17/2010	0.0	23.5	58.8	E	1.1	0	Near neutral
Prairie Creek at Highway 99, near Grand Mound (12027540)	–	0.0	0	8/17/2010	–	–	–	–	–	–	–
Chehalis River below Prairie Creek, near Rochester (12027570)	278.0	–	8	8/17/2010	29.0	25.5	57.1	F	1.7	17.1	Gaining reach
Chehalis River above Scatter Creek, near Rochester (12027580)	298.0	–	8	8/17/2010	20.0	32.6	55.7	G	1.4	14.3	Near neutral
Scatter Creek at mouth, near Rochester (12028058)	–	16.1	8	8/17/2010	–	–	–	–	–	–	–
Chehalis River near Rochester (12028060)	308.0	–	8	8/17/2010	-6.1	34.3	54.2	H	1.5	-4.1	Near neutral
Chehalis River near Independence (12028080)	317.0	–	5	8/18/2010	9.0	29.3	52.5	I	1.7	[1]5.3	Near neutral
Independence Creek at Garrad Creek Road, near Oakville (12028105)	–	0.5	11	8/19/2010	–	–	–	–	–	–	–
Chehalis River below Independence Creek, near Independence (12028260)	329.0	–	5	8/18/2010	11.5	22.8	51.5	J	1.0	11.5	Near neutral
Chehalis River at river mile 50.2, near Oakville (12028270)	306.0	–	5	8/18/2010	-23.0	22.5	50.2	K	1.3	-17.7	Losing reach

Table 3. Evaluation of gains and losses, with error analysis measurements for the seepage investigation, central Chehalis River Basin, southwestern Washington, August 2010.—Continued

[Associated measurement error for net gain or loss was calculated using the propagation of error formula (Wheeler and Eddy-Miller, 2005) where s is the error propagated from all estimated individual errors and a, b, ..., n are estimated errors for the discharge measurement at each site: $s=\sqrt{((\pm a)\,x^2+(\pm b)\,x^2+...(\pm n)\,x^2)}$. Seepage reach: Seepage reach delineated in figure 9. **Remarks:** Near neutral— Difference in measured discharge is less than associated measurement error. ft³/s, cubic feet per second; (ft³/s)/mi, cubic feet per second per mile; mi, mile; –, no data]

Measurement sites and No.	Measured streamflow (ft³/s)		Assumed measurement error (percent)	Date	Net gain or loss (ft³/s)	Associated measurement error (ft³/s) +/-	Chehalis river mile (mi)	Seepage reach	Seepage reach length (mi)	Net gain or loss per river length [(ft³/s)/mi]	Remarks
	Mainstem	Tributary									
Chehalis River at river mile 49.1, near Oakville (12028280)	314.0	–	5	8/18/2010	8.0	21.9	49.1	L	11	7.3	Near neutral
Chehalis RIver at river mile 47.9, near Oakville (12028290)	316.0	–	8	8/18/2010	2.0	29.8	47.9	M	12	1.7	Near neutral
Chehalis River above Black River, near Oakville (12028300)	307.0	–	5	8/18/2010	-9.0	29.6	47.1	N	0.8	-11.3	Near neutral
Black River above mouth, near Oakville (12029220)	–	93.3	8	8/18/2010	–	–	–	–	–	–	–
Chehalis River at river mile 46.1, near Oakville (12029250)	352.0	–	5	8/19/2010	-48.3	24.5	46.1	O	1.0	[1]-48.3	Losing reach
Chehalis River above Garrard Creek, near Oakville (12029300)	386.0	–	8	8/19/2010	34.0	35.5	45.0	P	11	30.9	Near neutral
Garrard Creek near Oakville (12029500)	–	3.3	8	8/18/2010	–	–	–	–	–	–	–
Chehalis River near Oakville (12029700)	396.0	–	5	8/19/2010	6.7	36.7	44.0	Q	1.0	-6.7	Near neutral
Chehalis River near Cedarville (12029800)	403.0	–	5	8/18/2010	7.0	28.3	42.3	R	1.7	[1]4.1	Near neutral

[1] Upstream and downstream ends of seepage reaches were measured on different days.

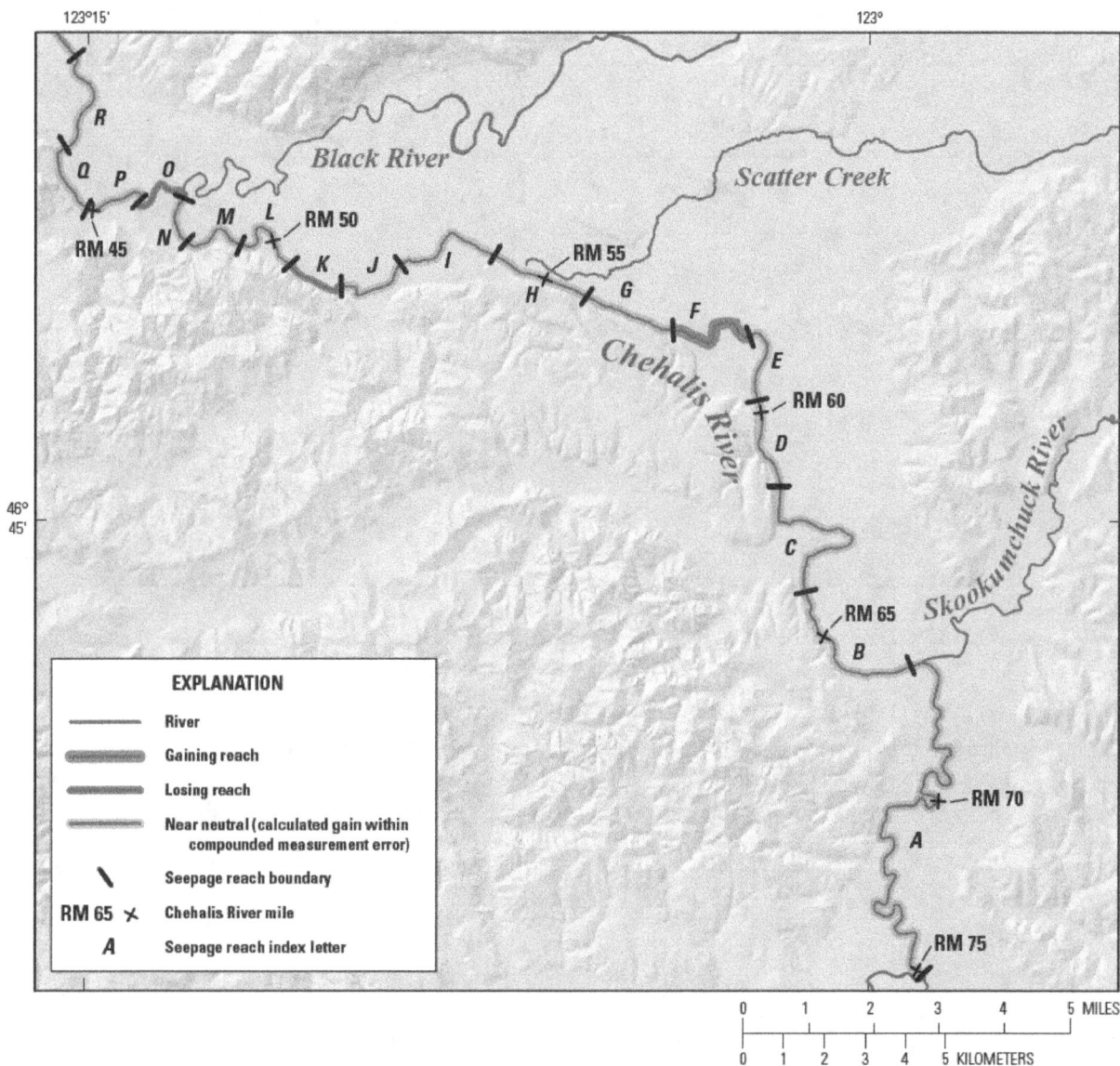

Figure 8.   Streamflow gaining, losing, and near-neutral reaches, central Chehalis River Basin, southwestern Washington, August 2010.

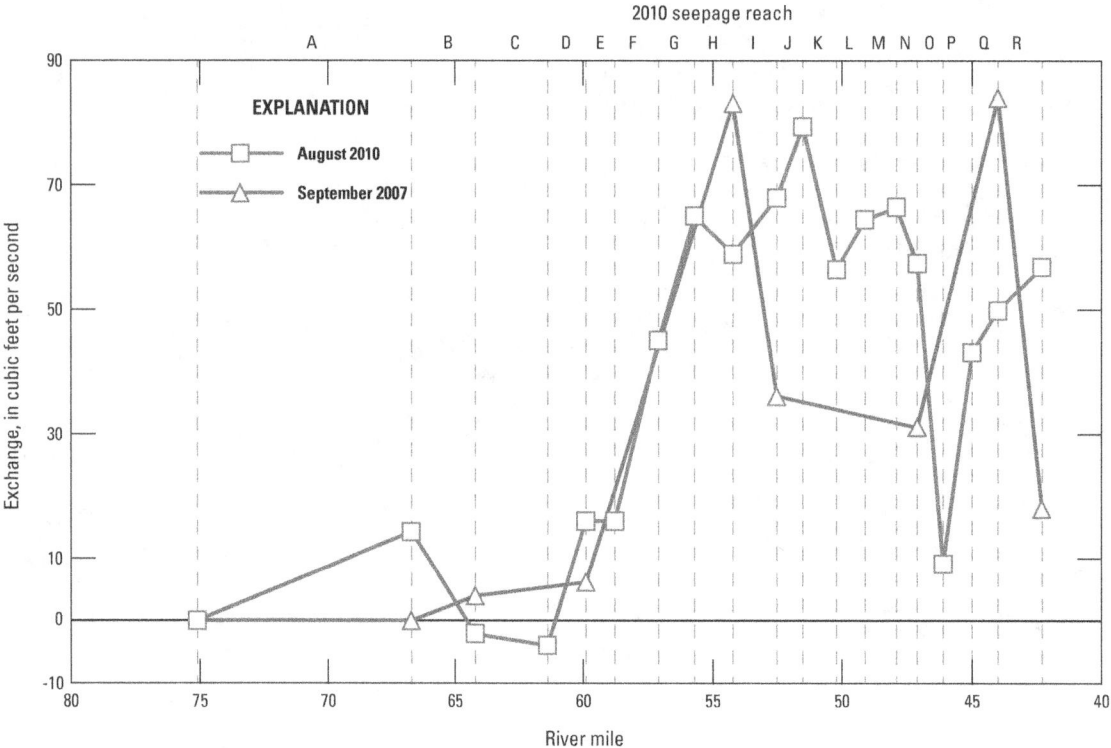

Figure 9. Cumulative river-aquifer exchanges for the central Chehalis River, southwestern Washington, August 2010 and September 2007.

The central Chehalis River flows over alluvial and glacio-fluvial sediments, which do not differ markedly in their hydraulic conductivity and therefore do not concentrate groundwater input along the Chehalis River. Finally, the significant tributaries that join the central Chehalis River, such as the Skookumchuck River, Scatter Creek, and the Black River flow over low-gradient topography formed during the last glacial retreat; the Chehalis River does not intercept any steep gradient tributaries that deposit alluvial that promote recharge of the main valley aquifer that could be subsequently intercepted by the mainstem Chehalis River.

Streamflow gains and losses also were estimated for several tributaries of the central Chehalis River including the Skookumchuck River, Scatter Creek, and Black River (table 4). Ungaged diversions and return flows between streamflow measurements may cause an overestimation of losses or gains, respectively. In particular, the reported gain of 6.5 (ft^3/s)/mi at the upstream-most seepage reach of Scatter Creek is likely the result of unmeasured hatchery return flows to Scatter Creek, which was inaccessible for measurement during the seepage run.

The groundwater levels within wells completed in the surficial aquifers near the Chehalis River fluctuated with the changes in river stage (fig. 10); note that groundwater levels and stream stage are not referenced to common datums). The changes in river stage were probably driven by precipitation events in the upper Chehalis River and tides within the lower Chehalis River. Wells completed within the unconfined A aquifer in close proximity to the Chehalis River and its tributaries, notably well 14N/02W-07B02, located at less than 0.1 mi from the Chehalis River, fluctuated directly with river stage (fig. 10A). Groundwater levels in well 14N/02W-07B02 responded to changes in gage height within several hours as

a result of the high hydraulic conductivities of the overlying sediments and lack of confining layers. The nearby well 14N/02W-06C02 (fig. 10B), which was also completed in the A aquifer fluctuates with the river stage, but the amplitude of the fluctuations was lower because it is further away (1.0 mi) from the Chehalis River. Similar relationships between river stage and groundwater levels in wells 14N/02W-07B02 and 14N/02W-06C02 were reported by Pitz and others (2005). Wells 15N/04W-02N03 (fig. 10C), 15N/04W-03R02 (fig. 10D) and 15N/03W-08B01 (fig. 10E) are completed in the A aquifer and show similarly muted groundwater-level responses to fluctuations in stream stage. Wells completed at greater distances away from the Chehalis River (for example, 15N/03W-03A02P3; 2.0 mi) have large seasonal fluctuations in water levels likely related to groundwater withdrawals during the summer and recharge from infiltrating precipitation during the fall and winter. Wells completed at greater distances from the Chehalis River have muted responses to fluctuations in river stage (fig. 10F) showing the attenuation of groundwater/surface-water interactions away from surface-water bodies.

The stage of the Chehalis River is influenced by ocean tides downstream of the confluence of the Chehalis River with the Satsop River as apparent at USGS streamflow gage 12035100. Water-level fluctuations of well 17N/07W-08K02 (fig. 11) completed in the A unit located less than 0.1 mi from the Chehalis River near its confluence with the Wynoochee River were similar to the tidally influenced fluctuations of the stage measured at a nearby USGS streamflow gage (12035100). Groundwater levels follow variations in tidally influenced river stage and are only attenuated several feet and delayed by several hours relative to the stream stage.

Table 4. Evaluation of gains and losses, with error analysis measurements for the seepage investigation tributaries of central Chehalis River, southwestern Washington, August 2010.

[Associated measurement error for net gain or loss was calculated using the propagation of error formula (Wheeler and Eddy-Miller, 2005) where s is the error propagated from all estimated individual errors and a, b, ..., n are estimated errors for the discharge measurement at each site: $s=\sqrt{((\pm a)\ x^2 + (\pm b)\ x^2 + ...(\pm n)\ x^2)}$. **Remarks:** Near neutral—Difference in measured discharge is less than associated measurement error. ft³/s, cubic feet per second; (ft³/s)/mi, cubic feet per second per mile; mi, mile; -, no data]

Measurement sites and No.	Measured streamflow (ft³/s)		Assumed measurement error (percent)	Date	Net gain or loss (ft³/s)	Associated measurement error (ft³/s)	Seepage reach length (mi)	Net gain or loss per river length [(ft³/s)/mi]	Remarks
	Mainstem	Tributary							
Skookumchuck River near Bucoda (12026400)	49.7	–	8	8/19/2010	–	–	–	–	–
Hanaford Creek at Hanaford Valley Road, near Centralia (12026580)	–	4.6	11	8/19/2010	–	–	–	–	–
Skookumchuck River at Centralia (12026600)	56.7	–	5	8/19/2010	2.4	4.9	4.3	0.6	Near neutral
Skookumchuck River at mouth, near Centralia (12026620)	73.4	–	8	8/17/2010	16.7	6.5	2.4	[1]7	Gaining reach
Scatter Creek at Case Road, near Grand Mound (12028020)	0.0	–	0	8/17/2010	–	–	–	–	–
Scatter Creek at Sargent Road, near Rochester (12028040)	20.8	–	8	8/18/2010	20.8	1.7	3.2	[1]6.5	Gaining reach
Scatter Creek at 183rd Avenue, near Rochester (12028043)	18.4	–	8	8/18/2010	-2.4	2.2	0.5	-4.8	Losing reach
Scatter Creek at Highway 12, near Rochester (12028050)	17.2	–	8	8/18/2010	-1.2	2.0	1.0	-1.2	Near neutral
Scatter Creek at James Road, near Rochester (12028055)	17.2	–	5	8/18/2010	0.0	1.6	0.8	0.0	Near neutral
Scatter Creek at Jordan Street, near Rochester (12028056)	18.8	–	5	8/18/2010	1.6	1.3	0.9	1.8	Gaining reach
Scatter Creek at mouth, near Rochester (12028058)	16.1	–	8	8/17/2010	-2.7	1.6	1.4	[1]-1.9	Losing reach
Black River at Gate (12029180)	62.3	–	11	8/18/2010	–	–	–	–	–
Black River near Oakville (12029200)	79.3	–	8	8/18/2010	17.0	9.3	3.2	5.3	Gaining reach
Willamette Creek near Howanut Road, near Oakville (12029201)	–	0.2	11	8/19/2010	–	–	–	–	–
Black River above mouth, near Oakville (12029220)	93.3	–	8	8/18/2010	13.8	9.8	4.3	3.2	Gaining reach

[1] Upstream and downstream ends of seepage reaches were measured on different days.

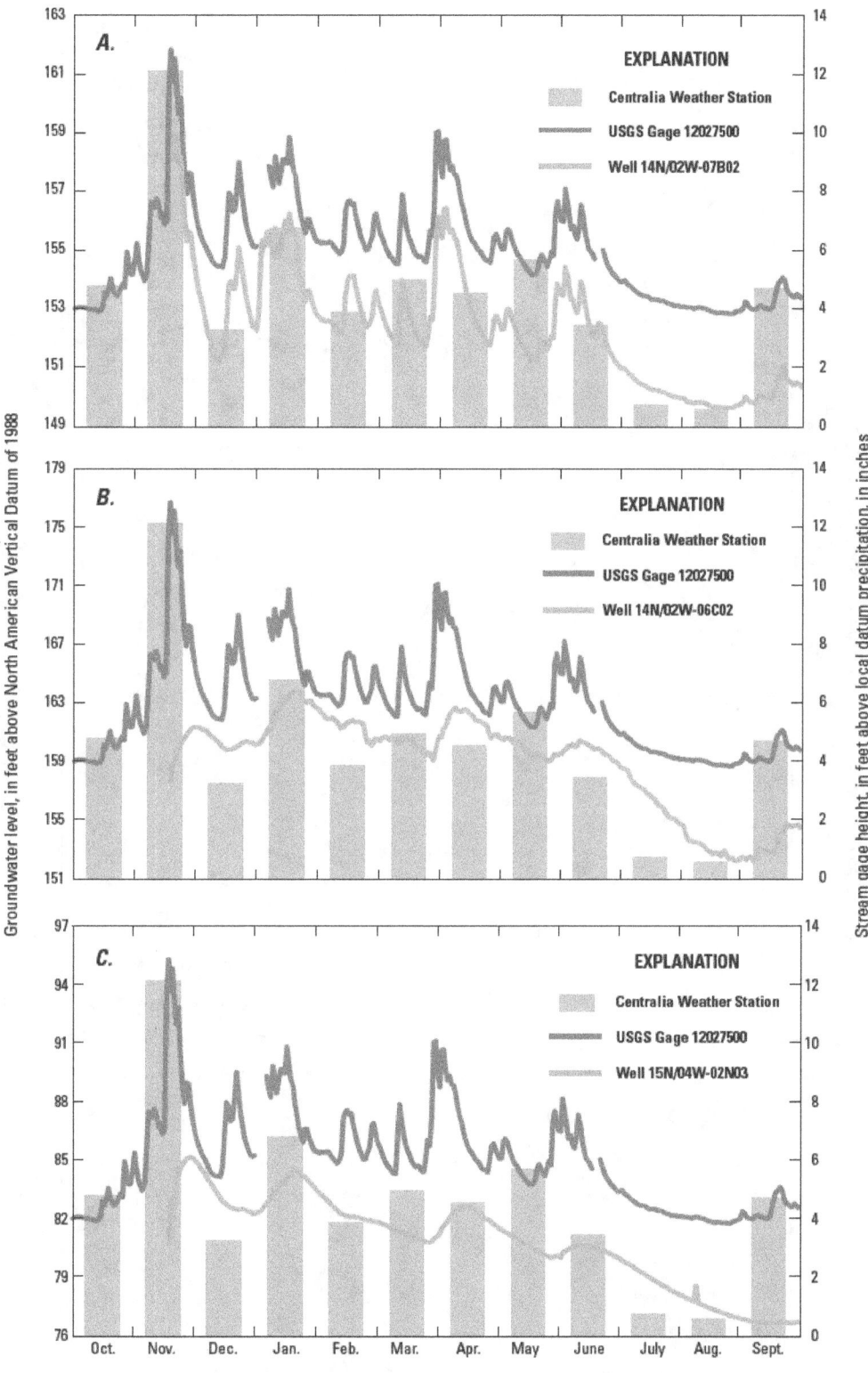

Figure 10. Continuous water levels in wells, stream stage at U.S. Geological Survey streamflow-gaging station 12027500, Chehalis River at Grand Mound, and total monthly precipitation at Centralia, October 2009–September 2010 for wells 14N/02W-07B02 (*A*), 14N/02W-06C02 (*B*), 15N/04W-02N03 (*C*), 15N/04W-03R02 (*D*), 15N/03W-08B01 (*E*), and 15N/03W-03A02P3 (*F*), Chehalis River basin, southwestern Washington.

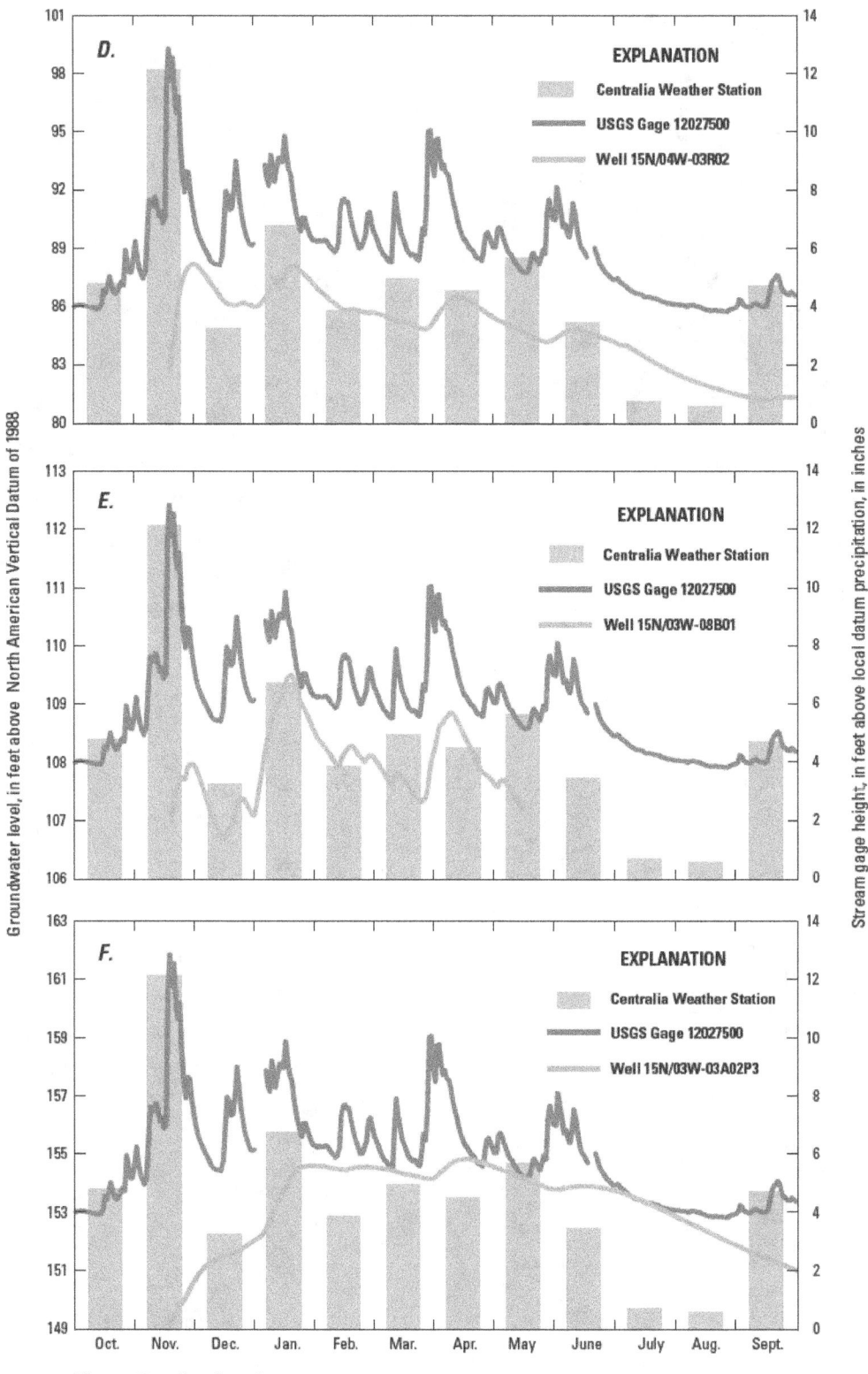

Figure 10.—Continued

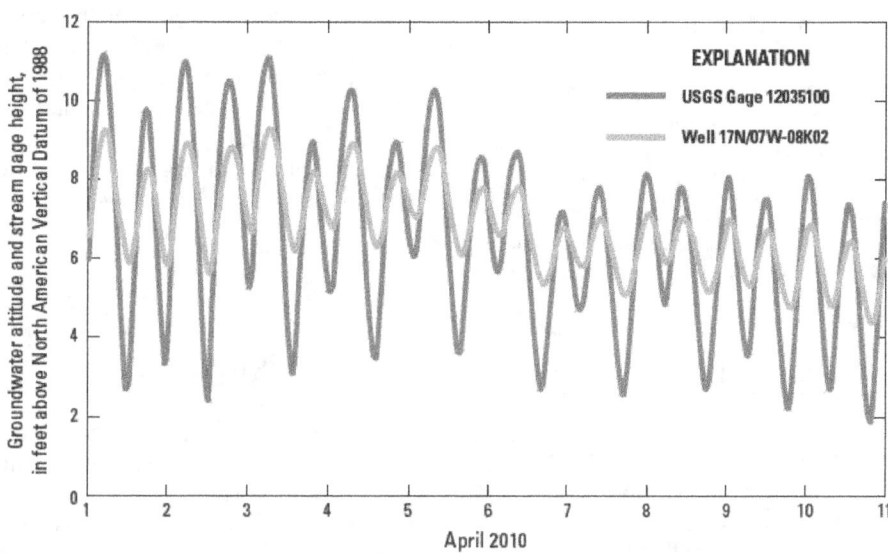

Figure 11.   Graph showing water levels in well 17N/07W-08K02 and stream stage at U.S. Geological Survey streamflow-gaging station 12035100, Chehalis River at Montesano, April 1, 2010–April 11, 2010, Chehalis River basin, southwestern Washington.

Summary and Conclusions

This study characterized the surficial hydrogeologic framework and groundwater/surface-water interactions within the 2,700-square-mile Chehalis River basin. The broad valleys and prairies of the Chehalis River basin are underlain by recent alluvium and glacial drift and contain most of the significant groundwater resources, agricultural lands, and urban development in the region. The Tertiary sedimentary and volcanic bedrock, which crops out at the surface along much of the Chehalis River basin uplands, has lower hydraulic conductivity than the alluvium and glacial drift and generally is not used as a source of water except for domestic consumption.

The unique position of the Chehalis River basin at the terminus of multiple advances of the Puget Lobe of the Cordilleran Ice Sheet resulted in a complex and heterogeneous deposition of sediments throughout the Quaternary Period. Aquifers within the outwash units deposited by meltwater streams that drained the Puget Lobe generally have high hydraulic conductivities. Glacial till, which generally has lower hydraulic conductivity, forms confining units which are often discontinuous because of post-glacial erosion and non-uniform deposition. Multiple advances of alpine glaciers originating in the Olympic and Cascade Mountains during the Pleistocene also deposited significant amounts of glacial outwash on the bedrock uplands of the eastern and northern parts of the Chehalis River basin.

The five hydrogeologic units and low permeability basal bedrock unit presented within this report are an attempt to generalize the Quaternary stratigraphy into a regional hydrogeologic framework, with the recognition that significant local variation exists. A surficial hydrogeologic unit map and two hydrogeologic cross-sections were constructed from existing geologic maps and information from 372 lithostratigraphic logs. Water levels measured during August–September 2009 were used to construct a water table map and generalized flow directions for the surficial aquifers of the Chehalis River basin.

Interaction between groundwater and surface water is apparent from water-level fluctuations in streams and aquifers and streamflow gains and losses measured during seepage runs. Synoptic streamflow measurements made in August 2010 show an overall gain of 56.8 ± 23.7 cubic feet per second over 32.8 river miles (1.7 cubic feet per second per mile) within the central Chehalis River and individual seepage reach gains and losses ranging from -48.3 to 30.9 cubic feet per second per mile. Most gains and losses, however, were less than compounded measurement error suggesting generally diffuse groundwater inflow, but became more pronounced downstream of Grand Mound. The uppermost unconfined aquifer (A unit) exchanges water with the Chehalis River and its tributaries, contributing baseflow during the dry summer months and receiving recharge from surface-water units during the wet winter months. Water levels in wells open to the A unit in close proximity to the surface-water features respond rapidly to changes in stream stage. Near-river wells respond most quickly to stream stage, but water levels also are influenced by other factors. Water levels within surficial aquifers contained by the upper Chehalis River basin respond to changes in stream stage due to storms, while Pacific Ocean tides influence streamflow stage and surficial aquifer water levels in the lower Chehalis River basin downstream of the Satsop River.

Acknowledgments

The author would like to thank the wellowners within the Chehalis River basin for providing access to their wells and the local, State and Tribal agencies who facilitated access to their well networks including the Washington State Department of Ecology and the Confederated Tribes of the Chehalis Reservation.

References Cited

Bear, J., 1979, Hydraulics of groundwater: New York, McGraw-Hill, 569 p.

Bretz, J.H., 1913, Glaciation of the Puget Sound region: Washington Geologic Survey Bulletin no. 8, 244 p.

Crandell, D.R., 1963, Paradise debris flow at Mount Rainier, Washington: U.S. Geological Survey Professional Paper 388A, 84 p.

Crandell, D.R., and Miller, R.D., 1974, Quaternary stratigraphy and extent of glaciation in the Mount Rainier region, Washington: U.S. Geological Survey Professional Paper 847, 59 p.

Drost, B.W., 2005, Quality-assurance plan for ground-water activities, U.S. Geological Survey, Washington Water Science Center: U.S. Geological Survey Open-File Report 05-1126, 27 p. (Also available at http://pubs.usgs.gov/of/2005/1126/.)

Drost, B.W., Turney, G.L., Dion, N.P., and Jones, M.A., 1998, Hydrology and quality of ground water in Northern Thurston County, Washington: U.S. Geological Survey Water-Resources Investigations Report 92-4109 [revised], 230 p. (Also available at http://pubs.er.usgs.gov/publication/wri924109_1994.)

Easterbrook, D.J., 1986, Stratigraphy and chronology of Quaternary deposits of the Puget Lowland and Olympic Mountains of Washington and the Cascade Mountains of Washington and Oregon: Quaternary Science Reviews, v. 5, p. 145-159.

Eddy, P.A., 1966, Preliminary investigation of the geology and groundwater resources of the lower Chehalis River valley and adjacent areas: Washington Department of Conservation Water Supply Bulletin 30, 70 p.

Ely, D.M., Frasl, K.E., Marshall, C.A., and Reed, F., 2008, Seepage investigation for selected river reaches in the Chehalis River basin, Washington: U.S. Geological Survey Scientific Investigations Report 2008-5180, 12 p. (Also available at http://pubs.usgs.gov/sir/2008/5180/.)

Fasser, E.T., and Julich, R.H., 2010, Groundwater levels for selected wells in the Chehalis River basin, Washington: U.S. Geological Survey Data Series 512. (Also available at http://pubs.usgs.gov/ds/512/.)

Ferris, J.G., Knowles, D.B., Brown, R.H., and Stallman, R.W., 1962, Theory of aquifer tests: U.S. Geological Survey Water-Supply Paper 1536-E, 174 p.

Garrigues, R.S., Sinclair, K., and Tooley, J., 1998, Chehalis River watershed surficial aquifer characterization: Washington State Department of Ecology, Publication No. 98-335, 33 p.

Heaton, T.H., and Kanamori, H., 1984, Seismic potential associated with subduction in the northwestern United States: Bulletin of the Seismological Society of America, v. 74, no. 3, p. 933 – 941.

Homer, C.C., Huang, C., Yang, L., Wylie, B., and Coan, M., 2004, Development of a 2001 national landcover database for the United States: Photogrammetric Engineering and Remote Sensing, v. 70, p. 829–840.

Konrad, C.P., 2006, Location and timing of river-aquifer exchanges in six tributaries to the Columbia River in the Pacific Northwest of the United States: Journal of Hydrology, v. 329, p. 444-470.

Lea, P.D., 1984, Pleistocene glaciation at the southern margin of the Puget Lobe, Western Washington: Master's thesis, University of Washington, 96 p.

Logan, R.L., compiler, 1987, Geologic map of the Chehalis River and Westport quadrangles, Washington: Washington Division of Geology and Earth Resources Open-File Report 87-8, 16 p., 1 plate, scale 1:100,000.

Logan, R.L., 2003, Geologic map of the Shelton 1:100,000 quadrangle, Washington: Washington Division of Geology and Earth Resources Open-File Report 2003-15, 1 sheet, scale 1:100,000.

Logan, R.L., Walsh, T.J., Stanton, B.W., and Sarikhan, I.Y., 2009, Geologic map of the Maytown 7.5-minute quadrangle, Thurston County, Washington: Washington Division of Geology and Earth Resources Geologic Map GM-72, 1 sheet, scale 1:24,000.

National Oceanic and Atmospheric Administration, 2010, Climatological data, annual summary - Washington, 2009: Ashville, North Carolina, National Climatic Data Center, v. 113, no. 13.

Noble, J.R., and Wallace, F.F., 1966. Geology and ground-water resources of Thurston County, Washington, v. 2: Washington Department of Conservation Water-Supply Bulletin 10, 61 p.

Oberg, K.A., Morlock, S.E., and Caldwell, W.S., 2005, Quality assurance plan for discharge measurements using acoustic Doppler current profilers: U.S. Geological Survey Scientific Investigations Report 2005-5183, 35 p. (Also available at http://pubs.usgs.gov/sir/2005/5183/.)

Pitz, C.F., and Sinclair, K.A, 1999, Estimated baseflow characteristics of selected Washington rivers and streams: Washington Department of Ecology Water Supply Bulletin No. 60, Publication 99-327, 25 p. + App.

Pitz, C.F., Sinclair, K.A., and Oestreich, A.J., 2005, Washington State groundwater assessment program— Hydrology and quality of groundwater in the Centralia– Chehalis area surficial aquifer: Washington State Department of Ecology, Publication no. 05-03-040, 104 p., 4 pls.

PRISM Group, 2011, Oregon State University, Mean annual precipitation from 1971 to 2000, accessed online at http://prism.oregonstate.edu/products/matrix.phtml on January 5, 2011.

Rantz, S.E., 1982, Measurement and computation of streamflow, volume 1—Measurement of stage and discharge: U.S. Geological Survey Water-Supply Paper 2175, 284 p. (Also available at http://pubs.usgs.gov/wsp/wsp2175/.)

Sauer, V.B., and Meyer, R.W., 1992, Determination of error in individual discharge measurements: U.S. Geological Survey Open-File Report 92–144, 21 p. (Also available at http://pubs.usgs.gov/of/1992/ofr92-144/.)

Schasse, H.W., compiler, 1987, Geologic map of the Centralia quadrangle, Washington: Washington Division of Geology and Earth Resources Open File Report 87-11, 28 p., 1 plate, scale 1:100,000.

Sinclair, K.A., and Hirschey, S.J., 1992, A hydrogeologic investigation of the Scatter Creek/Black River area, southern Thurston County: Master's thesis, Evergreen State College, 192 p.

Snavely, P.D. Jr., Brown, R.D. Jr., Roberts, A.E., and Rau, W.W., 1958, Geology and coal-resources of the Centralia-Chehalis district, Washington, with a section on microscopical character of the Centralia-Chehalis coal, by J.M. Schopf: U.S. Geological Survey Bulletin 1053, 159 p.

Thackray, G.D., 1996, Glaciation and neotectonic deformation on the western Olympic Peninsula, Washington: Phd. thesis, University of Washington, 139 p., 2 plates.

U.S. Geological Survey, 2011, Annual water-data reports: U.S. Geological Survey, accessed July 14, 2011 at http://wdr. water.usgs.gov/.

Washington Division of Geology and Earth Resources, 2005, Digital 1:100,000-scale geology of Washington State, version 1.0: Washington Division of Geology and Earth Resources Open-File Report 2005-3, accessed January 5, 2011 at http://www.dnr.wa.gov/ResearchScience/Topics/ GeosciencesData/Pages/gis_data.asp.

Wallace, E.F., and Molenaar, D., 1961, Geology and ground-water resources of Thurston County, Washington, Volume I: Washington Department of Conservation Water Supply Bulletin 10, 254 p.

Weigle, J.M., and Foxworthy, B.L., 1962, Geology and groundwater resources of west-central Lewis County, Washington: Washington Department of Conservation Water Supply Bulletin 17, 248 p.

Wells, R.E., Engebretson, D.C., Snavely, P.D. Jr., and Coe, R.S., Cenozoic plate motions and the volcanic-tectonic evolution of Oregon and Washington, Tectonics, v. 3, p. 275 – 294.

Wheeler, J.D. and Eddy-Miller, C.A., 2005, Seepage investigation on selected reaches of Fish Creek, Teton County, Wyoming, 2004: U.S. Geological Survey Scientific Investigations Report 2005-5133, 15 p. (Also available at http://pubs.usgs.gov/sir/2005/5133/.)

Table 5   29

Table 5. Select physical and hydrologic data for the project wells in the Chehalis River basin, southwestern Washington.

[Site No.: See well-numbering system diagram for explanation of well-numbering system. **Type of finish**: O, Open Hole; O, Open End; P, Perforated or Slotted; S, Screen; X, Open Hole. **Land-surface and water-level altitudes**: Referenced to the North American Vertical Datum of 1988 (NAVD 88). **Date of measurement**: Date of measurement during 2009 well inventory. **Status of water level**: B, tidally influenced; F, flowing; O, obstructed; P, pumping; R, recently pumped; T, nearby well in same aquifer pumping. **Latitude and longitude**: In degrees, minutes, seconds referenced to the North American Datum of 1983 (NAD83). **Hydrogeologic unit of open interval**: A, aquifer; B, confining unit; C, aquifer; D, aquifers and confining units; E, aquifer; BDRK, low permeability basal bedrock unit. **Remarks**: Monthly, manual monthly water-level measurements; Transducer, hourly water-level measurements measured with transducer; Previously inventoried, well inventoried during prior USGS study. ft-bls, feet below land surface; ft, feet; na, not applicable; –, no data]

Site No.	Depth of hole (ft-bls)	Type of finish	Land-surface altitude (ft)	Water-level altitude (ft)	Date of measurement	Status of water level	Latitude	Longitude	Hydrogeologic unit of open interval	Remarks
12N/02W-03A02	66	O	463	454	08-10-09	–	46°33'31"N	122°54'40"W	E	–
12N/04W-04B01	60	P	408	376	08-13-09	R	46°33'31"N	123°11'01"W	BDRK	–
13N/01E-31M01	40	S	456	439	08-06-09	–	46°33'55"N	122°44'15"W	A	–
13N/01E-33P01	110	X	707	669	07-28-09	R	46°33'41"N	122°41'29"W	E	–
13N/01E-08P01	unknown	–	560	537	08-05-09	–	46°37'11"N	122°50'06"W	No log	–
13N/01E-08P02	98	O	566	539	08-05-09	–	46°37'12"N	122°50'10"W	BDRK	–
13N/01E-08Q01	557	O	445	347	08-05-09	–	46°37'12"N	122°49'50"W	BDRK	–
13N/01E-09D02	114	O	574	548	08-05-09	–	46°37'52"N	122°49'24"W	E	–
13N/01E-10L01	224	O	323	na	08-05-09	F	46°37'26"N	122°47'38"W	BDRK	–
13N/01E-16D01	236	X	319	256	08-11-09	–	46°37'08"N	122°49'14"W	BDRK	–
13N/01E-16D02	110	–	318	277	08-11-09	–	46°37'06"N	122°49'17"W	No log	–
13N/01E-17C01	140	X	556	529	08-05-09	–	46°37'10"N	122°50'06"W	BDRK	–
13N/01E-17Q01	78	O	734	718	08-06-09	–	46°36'29"N	122°42'28"W	E	–
13N/01E-18P01	132	O	682	573	07-29-09	–	46°36'27"N	122°43'53"W	E	–
13N/01E-21H02	220	P	806	633	08-10-09	–	46°36'01"N	122°40'47"W	BDRK	–
13N/01E-21Q02	98	O	553	525	08-06-09	–	46°35'27"N	122°48'34"W	E	–
13N/01E-22M01	317	O	472	390	08-06-09	–	46°35'48"N	122°47'54"W	E	–
13N/01E-26D02	78	O	580	542	07-27-09	R	46°35'21"N	122°46'45"W	E	–
13N/01E-26K01	135	X	549	477	07-28-09	–	46°34'56"N	122°46'12"W	E	–
13N/01E-28M01	70	X	547	531	07-28-09	–	46°34'47"N	122°41'49"W	BDRK	–
13N/01E-30B02	41	O	325	306	08-03-09	–	46°35'25"N	122°51'03"W	A	–
13N/01E-31L02	30	O	457	447	07-28-09	–	46°34'04"N	122°44'04"W	A	–
13N/01E-32B02	130	X	546	na	07-29-09	O	46°34'24"N	122°42'18"W	BDRK	–
13N/01E-32L02	158	P	450	na	na	–	46°34'02"N	122°50'11"W	BDRK	–
13N/01E-32Q02	135	O	530	500	07-29-09	–	46°33'53"N	122°49'56"W	E	–
13N/01E-33G01	133.5	O	565	565	na	–	46°34'07"N	122°41'16"W	E	–
13N/01E-34H02	147	O	446	382	07-28-09	–	46°34'08"N	122°46'56"W	BDRK	–
13N/01E-38J01	59	X	304	na	na	–	46°35'48"N	122°51'17"W	A	–

Table 5. Select physical and hydrologic data for the project wells in the Chehalis River basin, southwestern Washington.—Continued

[**Site No.**: See well-numbering system diagram for explanation of well-numbering system. **Type of finish**: O, Open End; P, Perforated or Slotted; S, Screen; X, Open Hole. **Land-surface and water-level altitudes**: Referenced to the North American Vertical Datum of 1988 (NAVD 88). **Date of measurement**: Date of measurement during 2009 well inventory. **Status of water level**: B, tidally influenced; F, flowing; O, obstructed; P, pumping; R, recently pumped; T, nearby well in same aquifer pumping. **Latitude and longitude**: In degrees, minutes, seconds referenced to the North American Datum of 1983 (NAD83). **Hydrogeologic unit of open interval**: A, aquifer; B, confining unit; C, aquifer; D, aquifers and confining units; E, aquifer; BDRK, low permeability basal bedrock unit. **Remarks**: Monthly, manual monthly water-level measurements; Transducer, hourly water-level measurements measured with transducer; Previously inventoried, well inventoried during prior USGS study. ft-bls, feet below land surface; ft, feet, na, not applicable; –, no data]

Site No.	Depth of hole (ft-bls)	Type of finish	Land-surface altitude (ft)	Water-level altitude (ft)	Date of measurement	Status of water level	Latitude	Longitude	Hydrogeologic unit of open interval	Remarks
13N/02W-02A01	154	S	340	317	08-04-09	–	46°38'42"N	122°53'26"W	E	–
13N/02W-06K01	63	–	330	na	na	–	46°38'26"N	122°58'52"W	E	Previously inventoried
13N/02W-07Q02	118	P	360	na	na	–	46°37'19"N	122°58'50"W	E	Previously inventoried
13N/02W-08E04	396	P	1,070	928	08-06-09	–	46°37'35"N	122°35'25"W	BDRK	–
13N/02W-08J02	105	O	1,200	1,166	08-06-09	–	46°37'30"N	122°34'32"W	E	–
13N/02W-08N02	158	O	398	286	08-04-09	–	46°37'21"N	122°57'60"W	E	–
13N/02W-09E04	39	O	194	188	07-30-09	R	46°37'44"N	122°56'58"W	A	–
13N/02W-14C03	120	P	264	256	08-03-09	–	46°37'01"N	122°53'54"W	BDRK	–
13N/02W-15D01	60	X	220	214	08-11-09	–	46°37'01"N	122°55'30"W	A	–
13N/02W-16M01	95	O	372	329	08-11-09	–	46°36'33"N	122°56'53"W	E	–
13N/02W-18H02	112	O	381	301	08-12-09	–	46°36'49"N	122°58'32"W	E	–
13N/02W-19Q01	158	S	356	227	08-12-09	–	46°35'39"N	122°58'42"W	E	–
13N/02W-21Q02	146	O	456	323	08-04-09	–	46°35'32"N	122°56'07"W	E	–
13N/02W-24L01	57	S	295	285	08-04-09	–	46°35'41"N	122°52'54"W	BDRK	–
13N/02W-26K02	40	S	352	351	08-04-09	–	46°34'54"N	122°53'48"W	E	–
13N/02W-38H01	68	O	308	293	08-12-09	R	46°35'44"N	122°52'03"W	A	–
13N/03W-03D02	130	P	220	208	08-10-09	–	46°38'52"N	123°03'04"W	BDRK	–
13N/03W-04G02	120	P	241	213	07-30-09	–	46°38'36"N	123°03'43"W	BDRK	–
13N/03W-06J02	57	O	211	203	08-13-09	–	46°38'22"N	123°06'02"W	BDRK	–
13N/03W-06P01	180	P	214	201	08-12-09	–	46°38'04"N	123°06'32"W	BDRK	–
13N/03W-08J01	180	P	241	208	08-05-09	–	46°37'32"N	123°04'51"W	BDRK	–
13N/03W-09Q01	148	P	244	na	08-12-09	F	46°37'13"N	123°03'43"W	BDRK	–
13N/03W-10Q01	223	P	209	181	07-29-09	R	46°37'12"N	123°02'35"W	BDRK	–
13N/03W-11C02	63	O	193	164	07-30-09	–	46°38'01"N	123°01'38"W	A	–
13N/03W-14K01	198	O	404	234	08-10-09	–	46°36'38"N	123°01'22"W	E	–
13N/03W-15K01	160	P	371	274	08-10-09	–	46°36'42"N	123°02'37"W	E	–
13N/03W-15L01	155	P	407	277	08-10-09	R	46°36'42"N	123°02'42"W	BDRK	–
13N/03W-16H01	57	O	424	380	08-10-09	–	46°36'49"N	123°03'19"W	E	–
13N/03W-16K01	280	P	474	402	08-10-09	–	46°36'43"N	123°03'40"W	BDRK	–
13N/03W-18R01	220	P	640	548	08-05-09	–	46°36'28"N	123°05'52"W	BDRK	–

Table 5   31

Table 5. Select physical and hydrologic data for the project wells in the Chehalis River basin, southwestern Washington.—Continued

[Site No.: See well-numbering system diagram for explanation of well-numbering system. Type of finish: O, Open End; P, Perforated or Slotted; S, Screen; X, Open Hole. Land-surface and water-level altitudes: Referenced to the North American Vertical Datum of 1988 (NAVD 88). Date of measurement: Date of measurement during 2009 well inventory. Status of water level: B, tidally influenced; F, flowing; O, obstructed; P, pumping; R, recently pumped; T, nearby well in same aquifer pumping. Latitude and longitude: In degrees, minutes, seconds referenced to the North American Datum of 1983 (NAD83). Hydrogeologic unit of open interval: A, aquifer; B, confining unit; C, aquifer; D, aquifers and confining units; E, aquifer; BDRK, low permeability basal bedrock unit. Remarks: Monthly, manual monthly water-level measurements; Transducer, hourly water-level measurements measured with transducer; Previously inventoried, well inventoried during prior USGS study; ft-bls, feet below land surface; ft, feet; na, not applicable; –, no data]

Site No.	Depth of hole (ft-bls)	Type of finish	Land-surface altitude (ft)	Water-level altitude (ft)	Date of measurement	Status of water level	Latitude	Longitude	Hydrogeologic unit of open interval	Remarks
13N/03W-19A01	48	P	643	626	08-04-09	–	46°36'13"N	123°05'59"W	E	–
13N/03W-20D02	83	P	631	na	08-06-09	O	46°36'08"N	123°05'47"W	E	–
13N/03W-20E01	360	P	630	591	08-05-09	R	46°35'58"N	123°05'36"W	E	–
13N/03W-23A01	147	O	373	252	08-15-09	R	46°36'11"N	123°00'58"W	E	–
13N/03W-31K01	220	P	304	na	na	–	46°34'04"N	123°06'23"W	BDRK	–
13N/03W-31K02	70	P	332	na	na	–	46°34'04"N	123°06'23"W	BDRK	–
13N/04W-04R01	300	P	247	176	08-27-09	R	46°38'10"N	123°10'56"W	BDRK	–
13N/04W-04R02	126	–	245	227	08-27-09	–	46°38'10"N	123°10'50"W	No log	–
13N/04W-05Q01	100	P	274	255	08-27-09	–	46°38'15"N	123°12'25"W	BDRK	–
13N/04W-08B01	160	O	258	246	08-11-09	R	46°37'55"N	123°12'30"W	BDRK	–
13N/04W-08B02	204	–	252	236	08-11-09	–	46°37'56"N	123°12'35"W	No log	–
13N/04W-08B03	200	P	255	250	08-11-09	R	46°37'58"N	123°12'27"W	BDRK	–
13N/04W-10K01	70	P	253	231	08-11-09	–	46°37'28"N	123°10'02"W	BDRK	–
13N/04W-31Q02	340	P	751	705	08-13-09	R	46°33'48"N	123°13'49"W	BDRK	–
13N/04W-36A01	35	P	221	na	na	–	46°34'26"N	123°07'15"W	A	–
13N/05W-02L01	160	P	286	265	09-04-09	–	46°38'29"N	123°16'29"W	BDRK	–
13N/05W-02P02	83	P	310	297	08-28-09	–	46°38'06"N	123°16'34"W	BDRK	–
13N/05W-12C01	143	P	289	280	08-17-09	–	46°38'03"N	123°15'16"W	BDRK	–
13N/05W-26D01	143	X	377	371	09-03-09	–	46°35'20"N	123°16'53"W	A	–
13N/05W-26E02	20	O	381	373	08-28-09	–	46°35'15"N	123°16'51"W	A	–
14N/01W-23Q01	106	X	445	431	08-11-09	–	46°40'41"N	122°45'58"W	BDRK	–
14N/01W-26G01	73	P	434	430	08-11-09	–	46°40'24"N	122°46'02"W	BDRK	–
14N/01W-32F01	60	P	639	623	08-13-09	R	46°39'34"N	122°50'06"W	E	–
14N/02W-02R02	118	O	278	203	08-11-09	–	46°43'19"N	122°53'20"W	BDRK	–
14N/02W-06C02	92	P	183	151	08-10-09	–	46°44'03"N	122°59'04"W	A	Transducer
14N/02W-07B02	61	P	169	150	08-12-09	–	46°43'05"N	122°58'54"W	A	Transducer
14N/02W-11P01	174	P	383	315	08-11-09	–	46°42'28"N	122°53'55"W	E	–
14N/02W-13H01	97	P	491	421	08-11-09	–	46°42'00"N	122°52'14"W	BDRK	–
14N/02W-17G01	23	P	176	167	08-10-09	–	46°41'58"N	122°57'20"W	A	–
14N/02W-17G02	75	F	176	164	08-10-09	–	46°41'58"N	122°57'20"W	A	–

Table 5. Select physical and hydrologic data for the project wells in the Chehalis River basin, southwestern Washington.—Continued

[Site No.: See well-numbering system diagram for explanation of well-numbering system. **Type of finish**: O, Open End; P, Perforated or Slotted; S, Screen; X, Open Hole. **Land-surface and water-level altitudes**: Referenced to the North American Vertical Datum of 1988 (NAVD 88). **Date of measurement**: Date of measurement during 2009 well inventory. **Status of water level**: B, tidally influenced; F, flowing; O, obstructed; P, pumping; R, recently pumped; T, nearby well in same aquifer pumping. **Latitude and longitude**: In degrees, minutes, seconds referenced to the North American Datum of 1983 (NAD83). **Hydrogeologic unit of open interval**: A, aquifer; B, confining unit; C, aquifer; D, aquifers and confining units; E, aquifer; BDRK, low permeability basal bedrock unit. **Remarks**: Monthly, manual monthly water-level measurements; Transducer, hourly water-level measurements measured with transducer; Previously inventoried, well inventoried during prior USGS study. ft-bls, feet below land surface; ft, feet, na, not applicable; –, no data]

Site No.	Depth of hole (ft-bls)	Type of finish	Land-surface altitude (ft)	Water-level altitude (ft)	Date of measurement	Status of water level	Latitude	Longitude	Hydrogeologic unit of open interval	Remarks
14N/02W-17M02	28	F	173	166	08-10-09	–	46°41'50"N	122°58'04"W	A	–
14N/02W-17M03	19	F	173	166	08-10-09	–	46°41'50"N	122°58'04"W	A	–
14N/02W-17N01	64	P	172	164	08-10-09	–	46°41'44"N	122°58'04"W	A	–
14N/02W-18D02	60	P	256	240	07-28-09	–	46°42'10"N	122°59'23"W	BDRK	–
14N/02W-19H06	300	–	170	na	na	–	46°41'13"N	122°58'32"W	BDRK	Previously inventoried
14N/02W-23A03	41	O	218	210	08-11-09	–	46°41'23"N	122°53'21"W	BDRK	–
14N/02W-23M03	180	P	249	na	Na	–	46°40'57"N	122°54'20"W	BDRK	–
14N/02W-26K01	86	O	565	531	08-07-09	–	46°40'02"N	122°53'46"W	E	–
14N/02W-27L01	160	P	535	407	08-07-09	–	46°40'01"N	122°55'15"W	BDRK	–
14N/02W-28NO1	46	P	200	193	08-07-09	–	46°39'57"N	122°56'50"W	A	–
14N/02W-28Q05	160	P	216	209	08-07-09	–	46°39'52"N	122°56'12"W	BDRK	–
14N/02W-28Q06	unknown	–	229	225	08-07-09	–	46°39'53"N	122°56'10"W	No log	–
14N/02W-31C01	127	O	165	na	na	–	46°39'39"N	122°58'55"W	A	Previously inventoried
14N/02W-31P01	510	–	160	na	na	–	46°39'05"N	122°58'57"W	A	Previously inventoried
14N/02W-33E01	98	P	570	513	08-04-09	–	46°39'30"N	122°56'52"W	BDRK	–
14N/02W-34H04	308	–	330	283	07-27-09	R	46°39'23"N	122°54'48"W	BDRK	–
14N/02W-36C02	159	O	576	467	08-11-09	–	46°39'44"N	122°52'42"W	E	–
14N/02W-39A01	53	F	183	156	08-26-09	–	46°43'37"N	122°58'43"W	A	–
14N/02W-39L01	50	O	173	149	08-20-09	–	46°43'16"N	122°59'21"W	A	–
14N/02W-39R01	60	S	174	152	08-12-09	–	46°43'12"N	122°58'56"W	A	–
14N/02W-41A01	67	F	185	172	08-12-09	–	46°43'47"N	122°57'11"W	A	Monthly
14N/03W-11G05	139	P	370	279	08-17-09	–	46°42'52"N	123°01'23"W	BDRK	–
14N/03W-11G06	75	O	367	316	08-17-09	–	46°42'51"N	123°01'22"W	E	–
14N/03W-13D03	60	X	216	198	07-28-09	R	46°42'11"N	123°00'35"W	BDRK	–
14N/03W-13H01	163	P	222	171	07-28-09	–	46°42'00"N	122°59'36"W	BDRK	–
14N/03W-19D01	48	P	336	312	08-12-09	–	46°41'20"N	123°07'01"W	BDRK	–
14N/03W-19D02	138	X	336	na	na	–	46°41'19"N	123°07'00"W	BDRK	–
14N/03W-27D01	140	P	579	541	08-13-09	–	46°40'27"N	123°03'19"W	BDRK	–
14N/03W-27J01	67	P	486	471	08-12-09	–	46°40'03"N	123°02'17"W	E	–
14N/03W-28P01	80	X	489	456	07-30-09	–	46°39'54"N	123°04'13"W	E	–

Table 5   33

Table 5. Select physical and hydrologic data for the project wells in the Chehalis River basin, southwestern Washington.—Continued

[Site No.: See well-numbering system diagram for explanation of well-numbering system. **Type of finish**: O, Open End; P, Perforated or Slotted; S, Screen; X, Open Hole. **Land-surface and water-level altitudes**: Referenced to the North American Vertical Datum of 1988 (NAVD 88). **Date of measurement**: Date of measurement during 2009 well inventory. **Status of water level**: B, tidally influenced; F, flowing; O, obstructed; P, pumping; R, recently pumped; T, nearby well in same aquifer pumping. **Latitude and longitude**: In degrees, minutes, seconds referenced to the North American Datum of 1983 (NAD83). **Hydrogeologic unit of open interval**: A, aquifer; B, confining unit; C, aquifer; D, aquifers and confining units; E, aquifer; BDRK, low permeability basal bedrock unit. **Remarks**: Monthly, manual monthly water-level measurements; Transducer, hourly water-level measurements measured with transducer; Previously inventoried, well inventoried during prior USGS study. ft-bls, feet below land surface; ft, feet; na, not applicable; –, no data]

Site No.	Depth of hole (ft-bls)	Type of finish	Land-surface altitude (ft)	Water-level altitude (ft)	Date of measurement	Status of water level	Latitude	Longitude	Hydrogeologic unit of open interval	Remarks
14N/03W-35P01	400	P	326	203	08-12-09	–	46°39'02"N	123°01'28"W	BDRK	–
14N/04W-05D01	60	P	247	224	08-13-09	–	46°44'03"N	123°13'16"W	BDRK	–
14N/04W-13R01	101	P	336	287	08-12-09	–	46°41'43"N	123°07'20"W	BDRK	–
14N/04W-13R02	29.5	O	269	256	08-12-09	–	46°41'40"N	123°07'12"W	BDRK	–
14N/04W-15E01	60	P	351	309	07-14-09	R	46°42'12"N	123°10'48"W	BDRK	–
14N/04W-15F01	64	P	421	411	08-13-09	–	46°42'04"N	123°10'28"W	BDRK	–
14N/04W-15F02	unknown	–	315	226	08-13-09	T	46°42'08"N	123°10'23"W	No log	–
14N/04W-26Q01	100	P	278	272	08-13-09	–	46°40'00"N	123°08'57"W	BDRK	–
14N/04W-36L01	65	P	223	222	08-12-09	–	46°39'18"N	123°07'55"W	BDRK	–
14N/05W-12K01	95	P	345	331	08-13-09	–	46°42'44"N	123°15'18"W	BDRK	–
14N/05W-12K02	110	P	347	337	08-13-09	–	46°42'43"N	123°15'16"W	BDRK	–
15N/01W-03R01	78	O	194	162	08-17-09	–	46°48'38"N	122°59'48"W	D	–
15N/01W-07D02	87	S	250	233	08-27-09	–	46°48'13"N	122°51'52"W	D	–
15N/01W-27D01	80	P	250	209	08-13-09	R	46°45'43"N	122°48'08"W	BDRK	–
15N/02W-12A01	108	S	363	298	09-04-09	R	46°48'17"N	122°52'15"W	BDRK	–
15N/02W-18M01D1	400	P	223	27	08-27-09	–	46°47'03"N	122°59'31"W	BDRK	–
15N/02W-20C01	68	P	254	239	08-17-09	–	46°46'41"N	122°57'40"W	BDRK	–
15N/02W-26N02	79	P	254	196	08-13-09	R	46°45'01"N	122°54'15"W	BDRK	–
15N/02W-27M01	27	S	206	194	08-14-09	–	46°45'13"N	122°55'35"W	A	–
15N/02W-28D02	180	P	330	205	08-13-09	R	46°45'43"N	122°56'45"W	BDRK	–
15N/02W-28P01	142	S	197	181	08-12-09	–	46°45'11"N	122°56'27"W	A	–
15N/02W-32E01	140	S	470	392	08-20-09	R	46°44'38"N	122°58'11"W	E	–
15N/02W-32H03	58	O	199	188	08-13-09	–	46°44'44"N	122°57'07"W	A	–
15N/02W-34K02	380	P	491	343	08-14-09	R	46°44'21"N	122°54'58"W	BDRK	–
15N/02W-35D02	125	X	238	208	08-14-09	R	46°44'53"N	122°54'30"W	BDRK	–
15N/03W-02E02	80	S	178	145	08-11-09	–	46°49'00"N	123°01'59"W	D	–
15N/03W-02R01P1	99	P	182	157	08-26-09	–	46°48'29"N	123°01'01"W	D	–
15N/03W-02R01P2	68	P	182	157	08-26-09	–	46°48'29"N	123°01'01"W	D	–
15N/03W-02R01P3	34	P	182	157	08-26-09	–	46°48'29"N	123°01'01"W	D	–
15N/03W-03A02P1	125	P	181	148	08-26-09	–	46°49'17"N	123°02'17"W	D	Monthly
15N/03W-03A02P2	85	P	181	149	08-26-09	–	46°49'17"N	123°02'17"W	D	Monthly
15N/03W-03A02P3	52	P	181	149	08-26-09	–	46°49'17"N	123°02'17"W	D	Transducer

Table 5. Select physical and hydrologic data for the project wells in the Chehalis River basin, southwestern Washington.—Continued

[Site No.: See well-numbering system diagram for explanation of well-numbering system. Type of finish: O, Open End; P, Perforated or Slotted; S, Screen; X, Open Hole. Land-surface and water-level altitudes: Referenced to the North American Vertical Datum of 1988 (NAVD 88). Date of measurement: Date of measurement during 2009 well inventory. Status of water level: B, tidally influenced; F, flowing; O, obstructed; P, pumping; R, recently pumped; T, nearby well in same aquifer pumping. Latitude and longitude: In degrees, minutes, seconds referenced to the North American Datum of 1983 (NAD83). Hydrogeologic unit of open interval: A, aquifer; B, confining unit; C, aquifers and confining units; E, aquifer; D, aquifers and confining unit; BDRK, low permeability basal bedrock unit. Remarks: Monthly, manual monthly water-level measurements; Transducer, hourly water-level measurements measured with transducer; Previously inventoried, well inventoried during prior USGS study. ft-bls, feet below land surface; ft, feet; na, not applicable; –, no data]

Site No.	Depth of hole (ft-bls)	Type of finish	Land-surface altitude (ft)	Water-level altitude (ft)	Date of measurement	Status of water level	Latitude	Longitude	Hydrogeologic unit of open interval	Remarks
15N/03W-06A03	70	S	135	101	08-27-09	—	46°49'15"N	123°06'04"W	D	—
15N/03W-07J01	320	P	365	260	08-20-09	—	46°47'49"N	123°06'07"W	BDRK	—
15N/03W-07M01	100	P	128	118	08-19-09	—	46°47'53"N	123°07'03"W	BDRK	—
15N/03W-08B01	78	S	131	104	08-19-09	—	46°48'19"N	123°05'08"W	D	Transducer
15N/03W-08D02	58	O	128	na	na	—	46°48'24"N	123°05'43"W	A	—
15N/03W-10A06	unknown	—	174	142	09-02-09	—	46°48'23"N	123°02'14"W	No log	—
15N/03W-10C02	57	O	169	133	09-04-09	—	46°48'23"N	123°02'44"W	D	—
15N/03W-10D02P1	83	P	161	131	08-26-09	—	46°48'27"N	123°03'17"W	D	Monthly
15N/03W-10D02P2	60	P	161	131	08-26-09	—	46°48'27"N	123°03'17"W	D	Monthly
15N/03W-10D02P3	35	P	161	131	08-26-09	—	46°48'27"N	123°03'17"W	D	Monthly
15N/03W-10J01	50	O	169	141	08-17-09	—	46°47'53"N	123°02'07"W	D	Monthly
15N/03W-10Q04	78	S	150	na	na	—	46°47'43"N	123°02'28"W	A	—
15N/03W-12L01	77	P	185	156	08-17-09	—	46°48'01"N	123°00'13"W	D	—
15N/03W-12N03	62	O	181	157	08-13-09	—	46°47'49"N	123°00'46"W	A	—
15N/03W-13H01	50	S	170	153	08-13-09	—	46°47'11"N	122°59'50"W	D	—
15N/03W-14B04	unknown	—	166	139	08-20-09	—	46°47'33"N	123°01'12"W	No log	—
15N/03W-14E01	58	O	166	135	08-20-09	—	46°47'32"N	123°01'11"W	D	—
15N/03W-19A04	280	P	445	311	08-14-09	—	46°46'35"N	123°05'58"W	BDRK	—
15N/03W-20Q01	106	P	293	na	08-14-09	P	46°45'52"N	123°05'00"W	BDRK	—
15N/03W-23P01	33.5	F	141	131	08-26-09	—	46°46'05"N	123°01'29"W	A	—
15N/03W-24F04	70	F	163	na	na	—	46°46'30"N	123°00'19"W	D	—
15N/03W-24L01	45	O	155	na	na	—	46°46'09"N	123°00'26"W	D	Previously inventoried
15N/03W-25L08	59	O	163	140	08-14-09	—	46°45'20"N	123°00'14"W	D	—
15N/03W-27L01	200	P	374	271	08-17-09	—	46°45'22"N	123°02'56"W	BDRK	—
15N/03W-27L02	340	P	374	259	08-17-09	—	46°45'22"N	123°02'56"W	BDRK	—
15N/03W-27P02	120	P	383	309	08-17-09	—	46°45'12"N	123°02'49"W	BDRK	—
15N/03W-27P03	140	P	358	na	na	—	46°45'13"N	123°02'55"W	BDRK	—
15N/04W-02K01	39	G	104	88	08-13-09	—	46°48'53"N	123°08'41"W	A	—
15N/04W-02M05	39	P	109	77	08-13-09	—	46°48'43"N	123°09'19"W	A	—
15N/04W-02M06	39.75	P	112	80	08-13-09	—	46°48'49"N	123°09'26"W	A	—

Table 5   35

Table 5. Select physical and hydrologic data for the project wells in the Chehalis River basin, southwestern Washington.—Continued

[**Site No.**: See well-numbering system diagram for explanation of well-numbering system. **Type of finish**: O, Open Hole; P, Perforated or Slotted; S, Screen; X, Open Hole. **Land-surface and water-level altitudes**: Referenced to the North American Vertical Datum of 1988 (NAVD 88). **Date of measurement**: Date of measurement during 2009 well inventory. **Status of water level**: B, tidally influenced; F, flowing; O, obstructed; P, pumping; R, recently pumped; T, nearby well in same aquifer pumping. **Latitude and longitude**: In degrees, minutes, seconds referenced to the North American Datum of 1983 (NAD83). **Hydrogeologic unit of open interval**: A, aquifer; B, confining unit; C, aquifer; D, aquifers and confining units; E, aquifer; BDRK, low permeability basal bedrock unit. **Remarks**: Monthly, manual monthly water-level measurements; Transducer, hourly water-level measurements measured with transducer; Previously inventoried, well inventoried during prior USGS study. ft-bls, feet below land surface; ft, feet; na, not applicable; –, no data]

Site No.	Depth of hole (ft-bls)	Type of finish	Land-surface altitude (ft)	Water-level altitude (ft)	Date of measurement	Status of water level	Latitude	Longitude	Hydrogeologic unit of open interval	Remarks
15N/04W-02N03	38.25	P	107	78	11-17-09	–	46°48'42"N	123°09'16"W	A	Transducer
15N/04W-03K02	52	F	106	82	08-13-09	–	46°48'52"N	123°10'15"W	A	–
15N/04W-03P01	54	–	104	na	na	–	46°48'31"N	123°10'16"W	A	–
15N/04W-03R02	53	F	110	81	08-13-09	–	46°48'37"N	123°09'48"W	A	Transducer
15N/04W-04C02	unknown	O	83	70	08-13-09	–	46°49'10"N	123°11'42"W	No log	Monthly
15N/04W-12C02	39	O	105	94	09-04-09	–	46°48'23"N	123°07'59"W	A	Monthly
15N/04W-14L01	80	P	255	224	08-26-09	–	46°47'07"N	123°09'13"W	BDRK	–
15N/04W-14L02	280	P	257	167	08-26-09	–	46°47'08"N	123°09'09"W	BDRK	–
15N/04W-21B01	unknown	P	159	na	na	–	46°46'44"N	123°11'32"W	BDRK	–
15N/04W-21B02	100	P	163	124	08-27-09	–	46°46'43"N	123°11'34"W	BDRK	–
15N/04W-21D01	185	P	146	na	na	–	46°46'34"N	123°12'04"W	BDRK	–
15N/04W-24M01	76	O	513	497	08-14-09	R	46°46'06"N	123°08'12"W	BDRK	–
15N/04W-29C01	36	X	155	147	08-27-09	–	46°45'44"N	123°13'02"W	BDRK	–
15N/05W-01R01	200	X	95	72	08-14-09	–	46°48'41"N	123°15'01"W	BDRK	–
15N/05W-10A01	110	P	111	105	08-25-09	R	46°48'26"N	123°17'33"W	No log	–
15N/05W-11E01	77	P	166	na	na	–	46°48'04"N	123°17'08"W	BDRK	–
15N/05W-11H01	180	X	158	-4	08-27-09	R	46°48'08"N	123°16'01"W	BDRK	–
15N/05W-11M01	105	P	189	139	08-26-09	–	46°47'58"N	123°17'04"W	BDRK	–
15N/05W-12M01	204	P	189	36	08-27-09	R	46°48'03"N	123°15'56"W	BDRK	–
16N/01W-16L03	59	O	335	289	09-01-09	–	46°52'09"N	122°48'58"W	D	–
16N/01W-18F01	58	O	458	421	08-27-09	–	46°52'23"N	122°51'31"W	D	–
16N/01W-18F02	290	P	535	373	08-27-09	–	46°52'20"N	122°51'35"W	BDRK	–
16N/01W-28B01	445	O	456	310	09-01-09	–	46°51'01"N	122°48'36"W	BDRK	–
16N/01W-30G01	123	P	465	418	09-02-09	–	46°50'37"N	122°51'13"W	BDRK	–
16N/01W-30P01	260	P	368	243	09-01-09	R	46°50'16"N	122°51'29"W	BDRK	–
16N/01W-30P02	unknown	–	275	256	09-02-09	–	46°50'15"N	122°51'38"W	No log	–
16N/02W-02P01	380	P	201	189	08-12-09	–	46°53'45"N	122°56'58"W	BDRK	–
16N/02W-05D02	28.5	O	191	184	08-26-09	–	46°54'23"N	122°58'04"W	C	–
16N/02W-06M04	245	P	182	159	08-26-09	R	46°53'58"N	122°59'21"W	BDRK	–
16N/02W-06Q01	120	P	190	180	08-26-09	–	46°53'52"N	122°58'40"W	BDRK	–

Table 5. Select physical and hydrologic data for the project wells in the Chehalis River basin, southwestern Washington.—Continued

[Site No.: See well-numbering system diagram for explanation of well-numbering system. Type of finish: O, Open End; P, Perforated or Slotted; S, Screen; X, Open Hole. Land-surface and water-level altitudes: Referenced to the North American Vertical Datum of 1988 (NAVD 88). Date of measurement: Date of measurement during 2009 well inventory. Status of water level: B, tidally influenced; F, flowing; O, obstructed; P, pumping; R, recently pumped; T, nearby well in same aquifer pumping. Latitude and longitude: In degrees, minutes, seconds referenced to the North American Datum of 1983 (NAD83). Hydrogeologic unit of open interval: A, aquifer; B, confining unit; C, aquifer; D, aquifers and confining units; E, aquifer; BDRK, low permeability basal bedrock unit. Remarks: Monthly, manual monthly water-level measurements; Transducer, hourly water-level measurements measured with transducer; Previously inventoried, well inventoried during prior USGS study. ft-bls, feet below land surface; ft, feet; na, not applicable; –, no data]

Site No.	Depth of hole (ft-bls)	Type of finish	Land-surface altitude (ft)	Water-level altitude (ft)	Date of measurement	Status of water level	Latitude	Longitude	Hydrogeologic unit of open interval	Remarks
16N/02W-09D02	92	X	197	181	08-27-09	–	46°53'35"N	122°56'57"W	BDRK	–
16N/02W-09D03	260	X	200	159	08-27-09	R	46°53'35"N	122°56'57"W	BDRK	–
16N/02W-09F01	305	P	290	193	08-29-09	R	46°53'20"N	122°56'37"W	BDRK	–
16N/02W-09N01	440	P	282	na	na	–	46°52'53"N	122°56'51"W	BDRK	Previously inventoried
16N/02W-10C02	235	–	284	283	08-12-09	–	46°53'32"N	122°55'19"W	No log	–
16N/02W-13P02	59	O	432	382	08-27-09	–	46°52'04"N	122°52'54"W	C	–
16N/02W-14J02	164	O	342	221	08-27-09	–	46°52'11"N	122°53'16"W	D	–
16N/02W-15K02	110	O	277	205	08-27-09	–	46°52'19"N	122°55'06"W	C	–
16N/02W-18Q01	320	P	309	264	09-03-09	–	46°52'02"N	122°58'41"W	BDRK	–
16N/02W-18Q02	unknown	–	312	294	09-03-09	–	46°52'00"N	122°58'40"W	No log	–
16N/02W-18Q03	200	–	308	276	09-03-09	–	46°52'00"N	122°58'39"W	No log	–
16N/02W-21R04	140	P	262	208	08-31-09	–	46°51'05"N	122°55'54"W	BDRK	–
16N/02W-27R03	100	O	263	228	09-02-09	–	46°50'22"N	122°54'41"W	D	–
16N/02W-29G04	78	O	234	192	08-31-09	R	46°50'50"N	122°57'26"W	C	–
16N/02W-29L02P1	108	P	218	na	na	–	46°50'32"N	122°57'41"W	D	Previously inventoried
16N/02W-31H02	unknown	–	204	183	08-26-09	–	46°49'36"N	122°58'27"W	No log	–
16N/02W-32L03	78	O	196	159	09-03-09	–	46°49'45"N	122°59'42"W	A	–
16N/02W-40H01	50	O	227	217	09-02-09	–	46°50'03"N	122°55'35"W	C	–
16N/03W-04P01	300	P	231	178	09-01-09	–	46°53'41"N	123°04'00"W	BDRK	–
16N/03W-04Q04	55	O	224	193	09-01-09	–	46°53'41"N	123°03'54"W	A	–
16N/03W-10H04	130	O	209	115	09-02-09	–	46°53'24"N	123°02'23"W	A	–
16N/03W-10H05	138.5	O	210	112	09-03-09	–	46°53'21"N	123°02'09"W	A	–
16N/03W-13R01	272	O	383	272	09-03-09	–	46°52'08"N	122°59'37"W	D	–
16N/03W-14P01	53	O	172	144	09-02-09	–	46°52'03"N	123°01'36"W	D	–
16N/03W-29L03P1	103	P	139	102	08-26-09	–	46°50'36"N	123°05'12"W	A	–
16N/03W-29L03P2	75	P	139	102	08-26-09	–	46°50'36"N	123°05'12"W	A	–
16N/03W-29L03P3	49	P	139	102	08-26-09	–	46°50'36"N	123°05'12"W	A	–
16N/03W-32A05	73	O	154	105	09-02-09	–	46°50'06"N	123°04'36"W	A	–
16N/03W-33F05	104	S	158	na	na	–	46°49'52"N	123°04'04"W	D	Previously inventoried
16N/03W-33P01P1	93	P	158	114	08-26-09	–	46°49'32"N	123°04'14"W	D	Monthly

Table 5   37

Table 5. Select physical and hydrologic data for the project wells in the Chehalis River basin, southwestern Washington.—Continued

[Site No.: See well-numbering system diagram for explanation of well-numbering system. **Type of finish**: O, Open Hole; P, Perforated or Slotted; S, Screen; X, Open Hole. **Land-surface and water-level altitudes**: Referenced to the North American Vertical Datum of 1988 (NAVD 88). **Date of measurement**: Date of measurement during 2009 well inventory. **Status of water level**: B, tidally influenced; F, flowing; O, obstructed; P, pumping; R, recently pumped; T, nearby well in same aquifer pumping. **Latitude and longitude**: In degrees, minutes, seconds referenced to the North American Datum of 1983 (NAD83). **Hydrogeologic unit of open interval**: A, aquifer; B, confining unit; C, aquifer; D, aquifers and confining units; E, aquifer; BDRK, low permeability basal bedrock unit. **Remarks**: Monthly, manual monthly water-level measurements; Transducer, hourly water-level measurements measured with transducer; Previously inventoried, well inventoried during prior USGS study. ft-bls, feet below land surface; ft, feet; na, not applicable; —, no data]

Site No.	Depth of hole (ft-bls)	Type of finish	Land-surface altitude (ft)	Water-level altitude (ft)	Date of measurement	Status of water level	Latitude	Longitude	Hydrogeologic unit of open interval	Remarks
16N/03W-33P01P2	70	P	158	114	08-26-09	—	46°49'32"N	123°04'14"W	D	Monthly
16N/03W-33P01P3	46	P	158	114	08-26-09	—	46°49'32"N	123°04'14"W	D	Monthly
16N/03W-34M02	78	O	179	136	09-04-09	—	46°49'43"N	123°03'13"W	D	—
16N/03W-36J01	138	P	200	na	na	—	46°49'44"N	122°59'39"W	D	Previously inventoried
16N/03W-50L01	84	O	188	na	na	—	46°49'37"N	123°00'06"W	D	—
16N/04W-19F01	unknown	X	481	na	na	—	46°51'16"N	123°13'39"W	BDRK	—
16N/04W-19R01	366	P	479	369	08-26-09	—	46°51'17"N	123°13'40"W	BDRK	—
16N/04W-26H02	59	O	113	89	08-27-09	—	46°50'40"N	123°08'39"W	A	—
16N/04W-30N01	104	S	80	60	08-27-09	—	46°50'15"N	123°14'32"W	A	—
16N/04W-31B02	52	P	82	71	08-13-09	—	46°50'07"N	123°13'49"W	A	—
16N/04W-31B03	79	S	82	68	08-27-09	—	46°50'04"N	123°13'60"W	A	—
16N/05W-09R01	57	O	71	52	08-27-09	—	46°53'05"N	123°18'36"W	A	—
16N/05W-09R02	56	O	72	na	na	—	46°53'00"N	123°18'34"W	A	—
16N/05W-09R03	58	O	71	na	na	—	46°53'04"N	123°18'35"W	A	—
16N/05W-14H01	100	P	110	72	08-26-09	—	46°52'26"N	123°16'01"W	A	—
16N/05W-16R01	63	O	78	46	08-25-09	R	46°52'01"N	123°18'35"W	A	—
16N/05W-22A01	60	O	71	50	08-11-09	—	46°51'53"N	123°17'20"W	A	Monthly
17N/02W-15J09	51.5	O	204	186	09-01-09	—	46°57'32"N	122°54'41"W	A	—
17N/02W-20B08	78	S	192	176	08-31-09	—	46°57'06"N	122°57'33"W	C	—
17N/02W-22F03	66	O	198	na	na	—	46°56'50"N	122°55'18"W	C	Previously inventoried
17N/02W-22M04	78	O	204	189	09-01-09	—	46°56'40"N	122°55'29"W	C	—
17N/02W-28J06	250	P	201	184	09-01-09	—	46°55'39"N	122°56'01"W	BDRK	—
17N/02W-30D03	180	P	166	148	09-03-09	—	46°56'12"N	122°59'19"W	BDRK	—
17N/02W-30M02	70	S	187	159	08-12-09	R	46°55'50"N	122°59'23"W	D	—
17N/02W-33A03	56	O	197	190	09-02-09	—	46°55'23"N	122°55'53"W	A	—
17N/02W-33K02	41	S	205	na	na	—	46°54'58"N	122°56'12"W	C	Previously inventoried
17N/02W-34R02	97	S	205	195	09-02-09	—	46°54'34"N	122°54'34"W	A	—
17N/02W-35C07	73	S	266	240	08-27-09	—	46°55'13"N	122°53'54"W	C	—
17N/02W-35C08	306	P	265	51	08-27-09	—	46°55'16"N	122°54'00"W	BDRK	—
17N/02W-35C09	306	P	275	203	08-27-09	R	46°55'24"N	122°53'55"W	BDRK	—

Table 5. Select physical and hydrologic data for the project wells in the Chehalis River basin, southwestern Washington.—Continued

[**Site No.**: See well-numbering system diagram for explanation of well-numbering system. **Type of finish**: O, Open End; P, Perforated or Slotted; S, Screen; X, Open Hole. **Land-surface and water-level altitudes**: Referenced to the North American Vertical Datum of 1988 (NAVD 88). **Date of measurement**: Date of measurement during 2009 well inventory. **Status of water level**: B, tidally influenced; F, flowing; O, obstructed; P, pumping; R, recently pumped; T, nearby well in same aquifer pumping. **Latitude and longitude**: In degrees, minutes, seconds referenced to the North American Datum of 1983 (NAD83). **Hydrogeologic unit of open interval**: A, aquifer; B, confining unit; C, aquifer; D, aquifers and confining units; E, aquifer; BDRK, low permeability basal bedrock unit. **Remarks**: Monthly, manual monthly water-level measurements; Transducer, hourly water-level measurements measured with transducer; Previously inventoried, well inventoried during prior USGS study. ft-bls, feet below land surface; ft, feet; na, not applicable; —, no data]

Site No.	Depth of hole (ft-bls)	Type of finish	Land-surface altitude (ft)	Water-level altitude (ft)	Date of measurement	Status of water level	Latitude	Longitude	Hydrogeologic unit of open interval	Remarks
17N/02W-35C10	202	X	272	254	08-26-09	—	46°55'23"N	122°53'58"W	C	—
17N/03W-11G04	67	S	224	179	09-02-09	—	46°58'37"N	123°01'27"W	A	—
17N/03W-11M02	700	P	307	252	09-02-09	—	46°58'27"N	123°01'54"W	BDRK	—
17N/03W-22L02	140	O	317	241	09-04-09	R	46°56'45"N	123°02'55"W	No log	—
17N/03W-23Q04	140	O	223	141	09-02-09	—	46°56'31"N	123°01'26"W	D	—
17N/03W-25R08	39	S	171	160	08-31-09	—	46°55'36"N	122°59'36"W	C	—
17N/03W-27R03	79	S	175	150	09-03-09	—	46°55'32"N	123°02'11"W	C	—
17N/03W-35E04	80	O	157	135	09-02-09	—	46°55'04"N	123°01'57"W	C	—
17N/03W-35F01	60.6	O	146	130	09-01-09	—	46°55'07"N	123°01'31"W	C	—
17N/03W-36N03	30	S	163	153	09-01-09	—	46°54'41"N	123°00'30"W	C	—
17N/03W-36R03	99	S	238	172	09-01-09	—	46°54'39"N	122°59'44"W	C	—
17N/05W-03D01	71	P	162	153	09-01-09	—	46°59'43"N	123°18'10"W	BDRK	—
17N/05W-06Q01	70.5	S	75	34	08-26-09	—	46°58'59"N	123°21'25"W	A	—
17N/05W-09L01	38	O	103	99	09-01-09	—	46°58'27"N	123°19'05"W	A	—
17N/05W-09P01	100	P	143	79	09-01-09	R	46°58'11"N	123°19'13"W	BDRK	—
17N/05W-10N01	48.6	O	101	81	08-27-09	—	46°58'11"N	123°18'04"W	BDRK	—
17N/05W-15D01	103	X	101	na	na	—	46°58'00"N	123°18'16"W	BDRK	—
17N/05W-33E01	58	O	67	40	08-25-09	—	46°55'13"N	123°19'22"W	A	—
17N/05W-34L01	121	P	119	99	08-27-09	—	46°54'56"N	123°17'58"W	BDRK	—
17N/06W-03D01	59	X	28	11	08-18-09	—	46°59'34"N	123°25'33"W	A	—
17N/06W-07D01	61	P	23	4	08-21-09	—	46°58'46"N	123°29'26"W	A	—
17N/06W-08E01	100	O	103	45	08-20-09	—	46°58'29"N	123°28'13"W	BDRK	—
17N/06W-08E02	180	X	106	4	08-18-09	R	46°58'28"N	123°28'15"W	BDRK	—
17N/06W-10R01	150	O	160	149	08-18-09	—	46°58'08"N	123°24'40"W	BDRK	—
17N/06W-10R02	53.5	O	139	120	08-18-09	—	46°58'15"N	123°24'39"W	A	—
17N/06W-11N01	65	S	157	131	08-18-09	—	46°58'04"N	123°24'22"W	A	—
17N/06W-13Q01	58	S	115	102	08-26-09	—	46°57'22"N	123°22'41"W	A	—
17N/06W-24C01	120	P	92	na	na	—	46°57'07"N	123°22'52"W	BDRK	—
17N/07W-01D01	35	O	37	25	08-18-09	—	46°59'34"N	123°30'34"W	A	—
17N/07W-03H01	50	P	51	44	08-20-09	—	46°59'29"N	123°32'15"W	A	—
17N/07W-05G02	51.5	O	89	66	08-21-09	—	46°59'18"N	123°34'53"W	A	—
17N/07W-08G01	130	P	13	3	08-25-09	B	46°58'27"N	123°34'54"W	A	—

Table 5   39

Table 5. Select physical and hydrologic data for the project wells in the Chehalis River basin, southwestern Washington.—Continued

[Site No.: See well-numbering system diagram for explanation of well-numbering system. **Type of finish**: O, Open Hole; O, Open End; P, Perforated or Slotted; S, Screen; X, Open Hole. **Land-surface and water-level altitudes**: Referenced to the North American Vertical Datum of 1988 (NAVD 88). **Date of measurement**: Date of measurement during 2009 well inventory. **Status of water level**: B, tidally influenced; F, flowing; O, obstructed; P, pumping; R, recently pumped; T, nearby well in same aquifer pumping. **Latitude and longitude**: In degrees, minutes, seconds referenced to the North American Datum of 1983 (NAD83). **Hydrogeologic unit of open interval**: A, aquifer; B, confining unit; C, aquifer; D, aquifers and confining units; E, aquifer; BDRK, low permeability basal bedrock unit. **Remarks**: Monthly, manual monthly water-level measurements; Transducer, hourly water-level measurements measured with transducer; Previously inventoried, well inventoried during prior USGS study. ft-bls, feet below land surface; ft, feet; na, not applicable; –, no data]

Site No.	Depth of hole (ft-bls)	Type of finish	Land-surface altitude (ft)	Water-level altitude (ft)	Date of measurement	Status of water level	Latitude	Longitude	Hydrogeologic unit of open interval	Remarks
17N/07W-08K02	203	P	9	-4	08-25-09	B	46°58'13"N	123°34'59"W	A	Transducer
17N/07W-11R01	74	O	23	7	08-19-09	–	46°58'04"N	123°30'48"W	A	–
17N/07W-11R02	unknown	–	23	8	08-18-09	–	46°58'04"N	123°30'48"W	No log	–
17N/07W-16C01	173	X	63	31	08-20-09	–	46°57'50"N	123°33'57"W	BDRK	–
17N/07W-17H01	60	P	156	130	08-20-09	–	46°57'43"N	123°34'47"W	BDRK	–
17N/08W-02G01	39	O	25	9	08-18-09	–	46°59'21"N	123°38'30"W	A	–
17N/08W-02G02	41	O	25	10	08-18-09	–	46°59'22"N	123°38'31"W	A	–
17N/08W-14H01	154.3	S	22	10	08-11-09	–	46°57'43"N	123°38'15"W	A	–
17N/08W-17A01	45	–	103	69	09-03-09	R	46°57'46"N	123°42'02"W	No log	–
17N/08W-23G01D1	114	P	22	16	08-25-09	B	46°56'50"N	123°38'31"W	A	–
17N/09W-01Q01	139.5	O	25	na	na	–	46°58'50"N	123°44'36"W	A	–
17N/09W-07B01	10	F	20	14	08-11-09	–	46°58'41"N	123°51'02"W	A	–
17N/09W-07B02	10	F	20	17	08-11-09	–	46°58'40"N	123°51'02"W	A	–
17N/09W-07C01	10	F	21	18	08-11-09	–	46°58'37"N	123°51'07"W	A	–
17N/09W-14L01	15	P	25	18	08-25-09	–	46°57'22"N	123°46'22"W	A	–
18N/05W-01P01	78	S	336	321	09-03-09	–	47°04'20"N	123°15'36"W	A	–
18N/05W-07B01	41	O	177	175	08-24-09	–	47°03'56"N	123°21'23"W	A	–
18N/05W-20L01	120	X	128	123	08-24-09	–	47°01'51"N	123°20'28"W	BDRK	–
18N/05W-23B01	465	P	356	115	09-01-09	R	47°02'26"N	123°16'25"W	A	–
18N/05W-24B01	160	X	273	264	09-04-09	–	47°02'26"N	123°15'17"W	A	–
18N/05W-26L01	39	P	216	205	09-04-09	–	47°01'00"N	123°16'34"W	BDRK	–
18N/05W-29M01	42	S	136	122	09-01-09	–	47°01'07"N	123°20'44"W	A	–
18N/05W-30C01	140	O	139	105	08-24-09	–	47°01'30"N	123°21'39"W	BDRK	–
18N/05W-30P01	64	P	113	91	08-24-09	–	47°00'49"N	123°21'32"W	A	–
18N/05W-32M01	120	X	252	176	08-19-09	–	47°00'14"N	123°20'40"W	BDRK	–
18N/06W-28P01	120	X	71	38	08-20-09	–	47°00'48"N	123°26'42"W	BDRK	–
18N/06W-30K01	43	O	69	54	08-20-09	–	47°00'55"N	123°28'36"W	A	–
18N/06W-33C01	104	O	55	29	08-20-09	–	47°00'26"N	123°26'40"W	A	–
18N/06W-34B01	160	O	65	26	08-20-09	–	47°00'26"N	123°25'07"W	BDRK	–
18N/06W-34B02	80	O	65	27	08-20-09	–	47°00'27"N	123°25'07"W	BDRK	–
18N/06W-34B03	80	O	72	na	na	–	47°00'31"N	123°25'05"W	BDRK	–

Table 5.   Select physical and hydrologic data for the project wells in the Chehalis River basin, southwestern Washington.—Continued

[Site No.: See well-numbering system diagram for explanation of well-numbering system. **Type of finish**: O, Open End; P, Perforated or Slotted; S, Screen; X, Open Hole. **Land-surface and water-level altitudes**: Referenced to the North American Vertical Datum of 1988 (NAVD 88). **Date of measurement**: Date of measurement during 2009 well inventory. **Status of water level**: B, tidally influenced; F, flowing; O, obstructed; P, pumping; R, recently pumped; T, nearby well in same aquifer pumping. **Latitude and longitude**: In degrees, minutes, seconds referenced to the North American Datum of 1983 (NAD83). **Hydrogeologic unit of open interval**: A, aquifer; B, confining unit; C, aquifer; D, aquifers and confining units; E, aquifer; BDRK, low permeability basal bedrock unit. **Remarks**: Monthly, manual monthly water-level measurements; Transducer, hourly water-level measurements measured with transducer; Previously inventoried, well inventoried during prior USGS study. ft-bls, feet below land surface; ft, feet; na, not applicable; —, no data]

Site No.	Depth of hole (ft-bls)	Type of finish	Land-surface altitude (ft)	Water-level altitude (ft)	Date of measurement	Status of water level	Latitude	Longitude	Hydrogeologic unit of open interval	Remarks
18N/07W-11H01	179	P	180	132	08-17-09	—	47°03'47"N	123°30'54"W	A	—
18N/07W-14G01	60	S	83	73	08-19-09	—	47°02'54"N	123°30'59"W	BDRK	—
18N/07W-24D01	90	X	76	62	08-19-09	R	47°02'13"N	123°30'34"W	A	—
18N/07W-25F01	184	P	105	13	08-19-09	—	47°01'09"N	123°30'17"W	A	—
18N/07W-34K01	220	P	152	140	08-20-09	—	46°59'60"N	123°32'32"W	BDRK	—
18N/07W-34N01	342	P	192	117	08-20-09	R	46°59'55"N	123°33'04"W	BDRK	—
18N/07W-34R01	59	O	112	63	08-21-09	—	46°59'48"N	123°32'01"W	A	—
18N/07W-34R02	233	—	124	70	08-17-09	—	46°59'48"N	123°32'16"W	No log	—
18N/08W-05A01	200	X	128	90	08-25-09	—	47°04'40"N	123°41'54"W	BDRK	—
18N/08W-16F01	32	—	69	60	08-25-09	—	47°02'49"N	123°41'22"W	No Log	—
18N/08W-17I01	28	S	73	68	08-25-09	—	47°02'40"N	123°41'46"W	A	—
18N/08W-28F01	122	P	78	58	08-25-09	—	47°01'01"N	123°41'23"W	BDRK	—
18N/08W-28L01	46	P	67	49	08-25-09	—	47°00'53"N	123°41'19"W	BDRK	—
18N/08W-29A01	133	O	84	64	08-25-09	—	47°01'18"N	123°42'04"W	BDRK	—
18N/09W-05R01	—	—	111	84	08-19-09	R	47°04'02"N	123°49'10"W	No log	—
18N/09W-08B01	70	X	51	38	08-18-09	—	47°03'46"N	123°49'47"W	BDRK	—
18N/09W-10K01	60	O	119	80	08-18-09	R	47°03'29"N	123°47'04"W	E	—
18N/10W-08P01	84	S	105	81	08-19-09	—	47°03'13"N	123°57'23"W	E	—
18N/11W-15D01	103	O	9	-2	08-21-09	—	47°03'15"N	124°02'31"W	E	—
19N/05W-15L01	75	S	366	346	09-26-09	—	47°08'03"N	123°18'04"W	C	—
19N/06W-04N01	61	O	295	261	08-24-09	R	47°09'21"N	123°26'43"W	D	—
19N/06W-29G01	180	P	227	201	08-20-09	—	47°06'19"N	123°27'17"W	BDRK	—
19N/06W-30C01	38	O	190	173	08-21-09	—	47°06'40"N	123°28'58"W	D	—
19N/06W-30M01	220	P	199	na	na	—	47°06'07"N	123°29'19"W	BDRK	—
19N/06W-30N01	300	P	185	147	08-21-09	—	47°05'59"N	123°29'18"W	BDRK	—
19N/06W-30Q01	82	P	125	116	08-21-09	—	47°05'59"N	123°28'48"W	BDRK	—
19N/06W-32D01	267	P	124	na	na	—	47°05'44"N	123°27'55"W	BDRK	—
19N/07W-01P01	126	O	206	126	08-26-09	R	47°09'17"N	123°30'04"W	BDRK	—
19N/07W-12B01	56	O	204	181	08-26-09	—	47°09'05"N	123°29'44"W	D	—
19N/07W-33B01	180	P	134	108	08-19-09	R	47°05'39"N	123°33'23"W	BDRK	—
19N/07W-33H01	50	O	108	73	08-17-09	—	47°05'25"N	123°33'18"W	A	—

Table 5   41

Table 5. Select physical and hydrologic data for the project wells in the Chehalis River basin, southwestern Washington.—Continued

[Site No.: See well-numbering system diagram for explanation of well-numbering system. **Type of finish**: O, Open End; P, Perforated or Slotted; S, Screen; X; Open Hole. **Land-surface and water-level altitudes**: Referenced to the North American Vertical Datum of 1988 (NAVD 88). **Date of measurement**: Date of measurement during 2009 well inventory. **Status of water level**: B, tidally influenced; F, flowing; O, obstructed; P, pumping; R, recently pumped; T, nearby well in same aquifer pumping. **Latitude and longitude**: In degrees, minutes, seconds referenced to the North American Datum of 1983 (NAD83). **Hydrogeologic unit of open interval**: A, aquifer; B, confining unit; C, aquifer; D, aquifers and confining units; E, aquifer; BDRK, low permeability basal bedrock unit. **Remarks**: Monthly, manual monthly water-level measurements; Transducer, hourly water-level measurements measured with transducer; Previously inventoried, well inventoried during prior USGS study. ft-bls, feet below land surface; ft, feet; na, not applicable; –, no data]

Site No.	Depth of hole (ft-bls)	Type of finish	Land-surface altitude (ft)	Water-level altitude (ft)	Date of measurement	Status of water level	Latitude	Longitude	Hydrogeologic unit of open interval	Remarks
19N/09W-22J01	193	O	217	40	08-18-09	–	47°06'49"N	123°46'39"W	E	–
20N/06W-10F01	65	O	432	394	08-24-09	–	47°14'07"N	123°25'31"W	D	–
20N/06W-11D01	232	O	458	404	08-26-09	–	47°14'20"N	123°24'28"W	BDRK	–
20N/06W-28C01	160	P	357	na	08-23-09	O	47°11'48"N	123°26'44"W	BDRK	–
20N/06W-28C02D1	200	P	357	343	08-26-09	–	47°11'48"N	123°26'44"W	BDRK	–
20N/06W-28F01	111	P	352	na	na	–	47°11'41"N	123°26'43"W	A	–
20N/10W-36L02	65	O	167	148	08-19-09	–	47°10'25"N	123°52'19"W	E	–
21N/08W-07D01	124	O	170	71	08-18-09	–	47°07'59"N	123°45'05"W	E	–